ECOLOGICAL RESTORATION
AND
POWER AT NIAGARA FALLS

P. M. Eckel

~ Botanical Services, St. Louis ~

Patricia M. Eckel, F.L.S.

Research Scientist, Missouri Botanical Garden
P.O. Box 299, St. Louis, MO 63166-0299 U.S.A.
Email: patricia.eckel@mobot.org

ISBN-9781530017737

This book is written in recognition of the contributions of

Andrew Haswell Green

in preserving the integrity of New York State parks.

The Niagara Reservation, later designated the Niagara Falls State Park, was established in 1885. It is the oldest continuously operating State Park in the United States. It is also the first State Park created under the power of eminent domain.

On May 23, 1963, the Niagara Reservation was declared a

United States National Historic Landmark

The illustration on the front cover of this book is an image of the decommissioned 1905 Ontario Power Company Generating Station on the Ontario side quite near the Horseshoe Falls, which is on the left. The Ontario Power Company was an American-owned firm. The station was decommissioned in 1999. The vacant facility remains in full view of visitors to the Falls.

On the back cover is an image of the grassy slope leading down to Luna Island from Goat Island. This is not the primeval forest called for by the original legislation establishing the Niagara Reservation.

TABLE OF CONTENTS

INTRODUCTION

This book documents, largely from primary sources, the political and economic climate that underlain the development of the New York State Reservation at Niagara Falls, and closely associated power generation. A detailed account of the attempts at ecological restoration of the floral heritage of islands at the brink of Niagara Falls, particularly of Goat Island, was published recently (Eckel, 2013). The present work focuses on the activities of the business interests that developed power at the Falls, and of the Commissioners of the Reservation who were in charge of "Preservation of the Scenery of Niagara Falls" (Report of the New York State Survey for 1879).

The "Botanical Heritage of Islands at the Brink of Niagara Falls (Eckel, 2013) deals with aspects of the physical setting (geology, hydrology, climate, etc.), pre-1885 land use, forest condition at the time the Reservation was established, vegetation in detail for particular potions of the Island, history of floral exploration, and an annotated list of the species known for the Island including a summary of the rare species. The "Botanical Heritage" is a historical analysis, showing changes in the floral composition over time. Certain chapters touch on the activities of the Commissioners of the Reservation, particularly the Introduction and the Vegetation sections.

The present volume is a companion to the "Botanical Heritage," and evaluates the influence of the Commissioners on the vegetation of both Goat Island and Prospect Point. The Commissioners were charged by the original legislation with restoring the beautiful primeval scenery of the Falls for the delight of visitors. Of importance is the ongoing relationship of the power companies at the Falls to the Commissioners' activities.

THE PICTORIAL TRADITION

The use of old representations and depictions to make inferences about past natural conditions and character at Niagara Falls is not a new idea. For example, James Hall, the New York State Geologist, used such images in a geological report on Niagara Falls (Hall, 1882) describing the fact of recession at the cataracts of Niagara:

> "All the historical evidence that we possess upon the subject [of recession] proves the falls to have receded: and, although there have been no monuments established, yet the representations of early travelers, when compared with the present condition of the falls, proves that a change has taken place, though we cannot be certain of its precise amount."

Hall produced a facsimile of a print of the falls seen from the west side of the gorge of the original published in F. Louis Hennepin's work (1697). In this print Hall noticed a "cross-fall" flowing from west to east from what is now the Canadian shore. The print "represents a projecting rock upon the west side of the river which turned a part of the water across the main fall, as seen in the sketch." This waterfall did not exist in the later nineteenth century. Luna Island is also not drawn, or did not exist to be drawn in this print, although Hall seems to excuse this due to the inaccessibility of Goat Island at that time. The text of Hennepin also describes this third falls, flowing from west to east.

Hall also mentioned the visit of Peter Kalm, student of Linnaeus, in 1750, who visited the falls seventy-two years after Hennepin. Kalm "distinctly alludes to the projecting rock, which forced the water out of its direct course, causing it to fall across the great fall. He speaks of this rock having fallen down a few years previous, and in his view of the falls the spot is indicated. In this interval of seventy years we find that the recorded observations of these two travelers prove precisely the same kind of change to have taken place, as we suppose to have occurred previously, and which has subsequently altered the outline and position of the falls."

The physiognomy or structural appearance of plant communities can be seen in broad outlines from these early depictions at Niagara Falls. The density of plant communities can be estimated. Various artists will, as in written accounts, provide more or less information of greater or lesser accuracy in their depictions, depending on the interest, skill and intent of the individual. To some extent a "formula" or iconography was established which became a part of the tradition of illustrating Niagara Falls, which was expected in any drawing, and with which draftsmen could make quick illustrations for books, magazines, popular literature, signage, newspapers, etc. Other artists, desiring some originality or wishing to imprint their own personality in their paintings offer differing views, different sorts of detail. Other artists again, pursuing the perfection of a certain style, such as the landscape artists of the Hudson School of painting, especially as they chose to represent their views with realism, provide ever more detail and standards in accuracy by which less exact illustrations could be compared. One extraordinary example of realistic painting are the sketches and paintings of Frederick Church, around 1857.

Conifer vegetation — Among the various cautions to be made in ascribing factual weight to aspects of any drawing of the vicinity of the cataracts, is the apparent practice of taking sketches or earlier depictions, especially those used as illustrations, and altering them, usually back in Europe in the earliest pictures, to improve, clarify or otherwise enhance the information provided by the earlier sketch. The tendency in this sort of practice is to reduce the visual complexity (and accuracy) of the view, to simplify and embolden the image. Another is to remove "blemishes" such as blasted trees, de-vegetated areas, and the like. Occasionally it appears that the conifer species (Pine and Hemlock) of the first drawing tends to resemble a European species, such as a Spruce, probably because the artists producing the second drawing were not the artist of the first, and were using imagery with which they were familiar.

One artist who did not "spruce up" or repair his Niagara trees was Thomas Davies, who painted Niagara from the Canadian prospect through Oak-conifer tree frames, showing native peoples. In the two paintings reproduced by Adamson (1985), headless tree-trunks are drawn, their crowns torn away. This kind of depiction gives some indication of either the senescence of the native trees, and/or mechanical stress on trees at the crestline. Blasted trees, their trunks ending in jagged stumps are also depicted in a drawing with unusual content among Niagara's illustrations done after a work by James Pattison Cockburn entitled "The Falls of Niagara." In "View of the Horse Shoe Falls from Goat Island, 1857" (Adamson, 1985, no. 109), one of the more elaborate picnics is drawn with several adult couples, children, pets and an individual seated on the ground drawing the view. A front-woods of a conifer is drawn directly facing the Falls behind which occur deciduous trees, at least four of which are without crowns. The smooth, clean trunks of deciduous trees are also shown on the ground, their fore-ends pointed down slope as though they were intentionally cast there before heaving over the crest onto the talus slope below.

This picture shows the peculiar conifer vegetation fronting Goat Island on this (southwest) side and the deciduous trees behind. An extensive shrub layer is shown in the background, not in the coniferous zone, but the deciduous. The artist made an attempt to emphasize the interaction of civilization and wilderness which may have been characteristic of many regional paintings of the time. One of the most striking icons of the civilizing of the North American wilderness is the presence of stumps as evidence of clearing in nineteenth century landscape portrayals. Although not a stump, one of the tree boles is conspicuously drawn to show its big sawn end. This raw surface is placed by the artist directly beside one corner of the picnic blanket indicting an association between their pleasure and the cutting of the forest.

It is obvious in this drawing that the forest had been cleared away to provide a greensward, or grassed and closely mown area on which picnics could be had. Such a lawn could not have existed here without personnel hired by the island owners to tend it. The crest woods had been stripped to provide a prospect out onto the river. This process of opening up the high bank forest may have created the condition of exposure by which the deciduous trees lost their crowns.

This particular depiction, and another after a drawing by the same artist (Adamson, 1985 no. 108, "The Falls of Niagara. View of Horseshoe Fall from Below Goat Island, 1857") are valuable because they give some evidence of conditions before the invention and general use of the camera. Prospect areas in the vicinity of the Falls were probably the first forested areas to be cleared. By the time photography came into general use, these areas probably looked totally different than they had originally, if for no other reason than the reality of trampling by thousands of visitors. The two depictions drawn after Cockburn's drawings "originally engraved in 1833" (Adamson, 1985), one of Porter's Bluff overlooking Terrapin Rocks, or what is now known as Terrapin Point, and the other of the base of the gorge probably near what was known as the American Landing, or the American Stairs, now the base of Prospect Point, show how dense the coniferous element was—almost certainly Eastern Hemlock (*Tsuga canadensis*), although this extraordinary density may be the result of later "touching up" of an earlier drawing. Later photographs of the wooded slope overlooking the Terrapin Rocks in the early years of this century, show a more open forest of younger trees dominated by deciduous trees with occasional conifers—

probably not the primitive condition. A whole series of early, pre-photography paintings and drawing show a dense conifer forest on the islands in the American channel, the island and mainland shorelines, crested areas of hills and gorge, and particularly at the very base of the talus slope at the bottom of the gorge. Loss of this forest today, and by the time photographs were widely used disguise the ecological response of the primitive forest to early lowered water levels in the Niagara River here when the area was above the falls, and the extent of what is probably now a rare herbaceous community established on the exposed riverbed.

Saw mills were some of the earliest human establishments at Niagara on both sides of the river, coinciding with the need of explorers and voyageurs for planks for ships, continuing into the military requirements of later history in the region, and culminating in one of the largest log, paper and pulp milling industries in New York State (Recknagel, 1923; Scott & Scott, 1983). Baring a biological cause for the early loss of this forest community at Niagara, such as inability to adapt to higher pollution regimes, the paper and related industries, now mostly gone with the depletion of the region's forest reserves, might be presumed a reason for the disappearance of this forest component (Recknagel, 1923).

Out of this treasury of representation, and compared with the visual characteristics of the present landscape at Niagara, images can be chosen that appear to accurately reflect the historic occurrence of certain tree species in their correct habitats and relationships. Certain icons, such as conifer trees drawn to resemble species of Spruce (*Picea* sp.) which never occurred naturally on Goat Island, as far as published reports by botanists ever related, can be dismissed as perhaps later renditions of sketches of native conifers made at the site, and perhaps painted out back in the region from which the artist had travelled, and with whose native species of conifer the artist was familiar. The valuable information to be had in this case was that there were conifers at Niagara.

Guessing the species of deciduous trees depicted in two centuries of drawings, is more hazardous than that of conifers, except where the bark is painted white and peeling, when Paper, or Canoe Birch (*Betula papyrifera*) is likely the species depicted, such as occur in the crest forests of the Niagara gorge, and which are frequent today on the crest of the gorge, especially on the American side.

Goat Island — An early engraving, "In the Woods at Goat Island" shows a tiny man with a walking stick on a path deep in a thick forest.

We are fortunate in that the crest forest is frequently drawn, since all of the famous prospects take place within a framework of trees. The crest is a relatively harsh place, in the face of the prevailing winds on the American side, and with shallow soils on the Canadian when not on the exposed crests of slopes. Conifers and birches are frequently painted, and Oaks are as frequent, and Staghorn Sumac occasional— primarily in early paintings where the observer is on the Canadian side looking across the gorge at Goat Island, as in paintings by Edward Hicks around 1830. Even Arbor Vitae (*Thuja occidentalis*) is sometimes drawn sprouting, with its characteristic bent-back growth form, out of fissures in the unlikely habitat of the dolomite caprock, easily seen in the gorge today.

General vegetation — A full account of the illustrated tradition at Niagara would make an extremely useful study in itself, but only the most general indications can be discussed here.

Kalm's facsimile print, used by Hall above, also showed an interesting distinction between a perimeter habitat on Goat Island, and a central habitat. The perimeter is of conifers, as is the bank of the river on the Canadian or southern shore, and conifers are present at the summit of a small hill on the mainland American or northern side. Conifers are also depicted on the western face of Goat Island, facing into what we know to be the prevailing winds and spray zone of the Horseshoe Falls—also at Prospect Point and the spray zone of the American Falls. Conifers are drawn on rocks in the cavity of the plunge pool, or the head of the gorge.

The primitive iconography of deciduous and evergreen trees is simple and without question—useful characteristics to employ in interpreting tree and habitat types from all sorts of illustrations depicting the falls: deciduous trees are drawn in oval or circular outline, evergreens in triangular, spiky outlines; deciduous trees are always lighter in color than the evergreens, and in autumn and winter scenes the contrast with the dark colorations of the conifers is unambiguous.

The coniferous possibilities are only four: White Pine (*Pinus strobus*), Hemlock (*Tsuga canadensis*), Arbor Vitae, or White Cedar (*Thuja occidentalis*), and Red Cedar (*Juniperus virginiana*) with a small likelihood of Balsam Fir (*Abies balsamea*), as this tree is and has been rare in the region (Zenkert, 1934) with a station reported for Niagara

Glen, several miles downstream at the base of the Niagara River gorge (Hamilton, 1943).

I have taken it that the vegetation in most depictions of the falls environment is not distorted beyond credibility through attempts at idealization or romaticization of nature so much as it is simplified to signs indicating tree species and community types. Most of the depictions consulted in the present study were made by people attempting to report back to an incredulous but skeptical and well-educated public a stupendous natural phenomena during times when the intended readers were experiencing a cultural explosion in the elucidation of natural history. They had an interest in being factual knowing full well, as the eighteenth and nineteenth century seemed to produce many travelogues, that their presentations would be tested sooner or later—as would their pictures. Their ability to accurately relate their experiences in one particular would relate to their credibility as a whole.

Also, the floristic representations are corroborated through time by different authors and painters, so diverse in time and origin as to be unlikely that they deliberately participated in a tradition based on authority rather than their own experience.

Vegetational characteristics seem to correspond to the distribution of species as they occur today, and as they are likely to occur—for example, today Arbor Vitae enjoys cold, wet boggy soil as may have occurred in the saturated soil margins on Goat Island, as drawn, and it occurs today in rocky bluffs in the dolomite caprock of the Niagara gorge, the talus below and the talus slopes in the plunge pool basin area at the head of the gorge—none of which contradicts early drawings on conifers in these areas.

It is probable that at least the shoreline visible around Goat Island was covered with Arbor Vitae because Francois Andre Michaux said "Goat Island ... is seen from the banks of the river to be bordered with Arbor Vitae" (1819). Wied-Neuwied (1843) indicated that "The shores of [Goat Island] are shaded by old pines and very large white cedars ...." As for the lower forest in the pit of the plunge pool area at the head of the gorge, Olmsted (1880) quoted Robinson (1895) in that the talus below is wooded "often so far below that you sometimes look from the upper brink down on the top of tall pines that seem diminished in size."

White Pines overtop the deciduous canopy of the primeval forest, and several old paintings show this contrast (for discussion see section on the central woods).

In Petrides' (1958) identification manual of trees and shrubs, in which silhouettes of various tree species are given, certain useful characteristics may be used to differentiate various conifer species. Generally, an attempt may be made to distinguish Red Cedar from Arbor Vitae, in that in the former, the branches may be higher off the ground, exposing the stem, than in the latter, where the branches descend almost to the ground, covering the stem or trunk. White Pine has graceful limbs extending horizontally with the extremities gently ascending. The long, bare trunk differentiated from a crown, and long plume-like needle-leaves growing in bunches of five and imparting a dense look to the limbs of the White Pine contrasts with the trunk covered with old or young limbs of the Hemlock, not displaying a distinct crown in its rather triangular silhouette. Hemlock presents a looser, tattered silhouette, with short needless, its branches not in the bold, heavy strokes of the Pine, but feathery, light tracery. In White Pine, the trunk may be seen in the crown as the needles are borne away from it out to the sunlight, but in Hemlock, the upper trunk is usually hidden in the foliage.

Habitat will also serve to separate Red Cedar from Arbor Vitae in photographs:

Red Cedar: intolerant of shade, open, sunny habitats, dry soils, isolated individuals, needles usually to the tree tops, successional: with weedy growth.

Arbor Vitae: tolerates shade, shaded habitats, wet soils [also along the calcareous rim rock where Red's might also grow], individuals clumped or contiguous, tree tops in old specimens frequently denuded, may be a climax species on river edges, with native growth.

One interesting study might be to examine a century of promotional and souvenir photographs of the Falls on postcards and other commercial printings to detect the relative distance of the observer in these pictures from the actual surface of the vegetation. It may be that with more and more destruction of the shrub communities, eradication of the herb and shrub layers in Goat Island's central woods and thinning of the trees, coupled with the establishment of greater lawn areas with interspersed shade trees, that the potential observer recedes further and further away from the landscape surface. Note that today nearly all promotional photographs are taken from airplanes and helicopters, or up in the tall towers on the Canadian shore, or from the Canadian prospect, or the vegetated landscape is blocked by a screen of falling water.

Aerial and other distance views distort the impression of the extent of vegetation on Goat Island. From above, the canopy in summer looks dense, close and forested, concealing the barren lawn-scapes beneath them. The intimate close-up views of the earlier part of the century used in the annual reports of the Commissioners to the New York State Legislature, for example, and early postcards, show a different habitat than now, and intimate views of vegetated areas are generally unavailable today.

Pictures, painted or photographed, constitute a rich resource for indirect evidence of the condition of the early environment at Niagara. Although the possibility of factual inaccuracy is granted, they provide a context in which to test hypotheses, suggest them, corroborate other testimony and generally give direction to reconstruction schemes of the early ecosystem and its appearance.

Examples of Early Imagery

There is a tremendous uncollected store of Niagara Falls imagery existing through the world today in the untold millions of tourist photographs alone held in millions of photograph albums and dusty attics, in archives of pictorial information, such as that of newspapers, not mention the works of art housed in historical and artistic institutions. The accumulation of visual images has gone on for three centuries. The present study has been only able to evaluate the earlier of such representations; surely additional details of importance can be learned from the array of photographs ultimately available.

Hennepin — Facsimile of the drawing of the Falls printed in Hennepin's narration, English edition, 1698 (Gardner, 1880). The cross fall is drawn on the right. The coniferous borders on the mainland and island are here first illustrated. Original in the John Carter Brown Library, Brown University.

Kalm — This drawing, which relies quite heavily on the previous one, was published in 1751 "based upon Kalm's description and maps of Popple and French prints" (Adams, 1927). The conifers are conspicuous on the bluffs and river margins. This clear zonation is continued in paintings of all kinds through the nineteenth century. Today, the forest composition of this area is so altered that this early plant community is completely gone, and can only be inferred from graphic and literary reference. The ladder down the face of Goat Island can be identified as a fancy, since the island was inaccessible until over 50 years after the depiction was printed. It may have been an attempt to illustrate and "Indian ladder," which was actually made from pine or Arbor Vitae trees made to lean against the caprock and the talus slope below. Gentleman's Magazine. London. February 1751 (Seibel, 1990).

Kalm — The same image as above but in a drawing without the ladder. Original in the John Carter Brown Library, Brown University.

Cockburn — The Falls of Niagara. View of the Horseshoe fall from below Goat Island, 1857, after a drawing by James Pattison Cockburn. Ackermann & Co., London. 1857. The two Cockburn drawings are very rich in factual detail. They must be seen full size and as close as possible to the original to make the best use of them. It is from full-sized prints of these two engravings that, of the possible conifers which can be attributed to the localities illustrated, that these trees are definitely Eastern Hemlock (*Tsuga canadensis*). The birches of the picture below are either Yellow or Canoe (*Betula lutea* or *B. papyrifera*), both of which grow in similar stations in the Niagara gorge.

Cockburn, second engraving — After a drawing my James Pattison Cockburn, Ackermann & Co., London, 1857. View of the Horseshoe Fall from Goat Island. Evidence from this and the preceding illustration is important in reconstructing the original forest composition, and suggesting objectives in a plan for restoring areas such as Terrapin Point.

Science — John James Barralet, Science Unveiling the Beauties of Nature to the Genius of America, 1814. New York Public Library, New York (in Adamson, 1985). Niagara Falls appears in the background of this allegorical drawing in the context of "a compendium of plants and animals unique to the new World" and alludes "to the active enterprise of botanists, zoologists, and ornithologists in America" (McKinsey, in Adamson, 1985). Niagara Falls is said to represent the "geological revelations of the new continent." As detailed by Eckel (2013), much field work in botany was conducted at Niagara Falls by a surprising number of early botanists. Natural history held the attention of the Victorian world, and Niagara Falls may be seen as a symbol of natural phenomena representative of North America. No one would have been more aware of the rhetoric of science, insofar as it related to reporting in the newspapers than Mark Twain. In his richly humorous essay on "The First Authentic Mention" included in the souvenir book (The Niagara Book, 1893), Twain writes a nested satire on intellectual awakening: of an adolescent young man introduced to matrimony through the story of Adam and Eve, of the aboriginal Indian and his cultural submersion in European culture. By doing so, he satirizes the "Free Niagara" rhetoric of Olmsted and other proponents of the Niagara Reservation: Adam calling Niagara the "Garden of Eden," but Eve "says it is all woods and rocks and scenery, and therefore has no resemblance to a garden. Says it looks like a park, and does not look like anything but a park. Consequently, without consulting me, it has been new-named—Niagara Falls Park . . . And already there is a sign up: Keep Off the Grass." When asked what the Falls were for, she says "they were only made for scenery—like the rhinoceros and the mastodon." More outrageous is his spoof of the zoological taxonomist, as Adam refuses to recognize the baby Cain as his little offspring and sees it as a species new to science— if only he can catch one. "If it dies, I will take it apart and see what its arrangements are. I never had a thing perplex me so." "As I discovered it, I have felt justified in securing the credit of the discovery by attaching my name to it, and hence have called it *Kangaroorum Adamiensis* . . . "since it resembles a kangaroo in its short front legs and long hind ones. The title of Twain's essay is also a spoof on the accumulation of early literary references of Niagara Falls that began to be published in the Annual Reports of the Commissioners and other documents, titles beginning with Hennepin. Twain, naturally, scoops all these bibliographers by discovering the "first authentic mention," and publishing "Extracts from Adam's Diary."

Church, Canadian View — The supreme example of objectivity in landscape painting are Frederick Church's studies of Niagara Falls. (Frederic Church, Niagara Falls, 1856. Private collection, in Adamson, 1985).

Church, American View — Frederick Church, ca. 1857. View of American Falls, Goat Island, and Horseshoe Falls in background. The illustration shown above gives, in detail the characteristics of the west end of Goat Island, with its sparsely vegetated talus slope and patches of bare sediments above on the bluffs (high bank) in the mid 1850's. Philips International Institute.

Farny — A view of Goat Island in the late 1800's, by Henry Farny. American Gallery, 19th Century. The west end, heavily vegetated.

Vanderlyn — A view showing conifer vegetation on Goat Island by John Vanderlyn (1775–1852), probably early 1800's. American Gallery, 19th Century.

Edwards — Generally, the best source of objective detail is from photographs such as this (Ernest Edwards, On the Three Sisters, ca. 1891, Buffalo and Erie County Historical Society, in Kowsky, 1985). Such a large boulder inside a forest setting occurs on the south side of the Second Sister Island. Note the conifers, now absent, and the tree roots exposed by surges in the River, probably during a past winter when such surges were frequent, and the strong winds which together, caused the trees to throw to the right. That this is on the south side of one of the islands is evident from the wind direction indicated by the reclining trees. Note the characteristic bark of Arbor Vitae and White Pine. A bit of soil compaction is also evident on the path in the lower right.

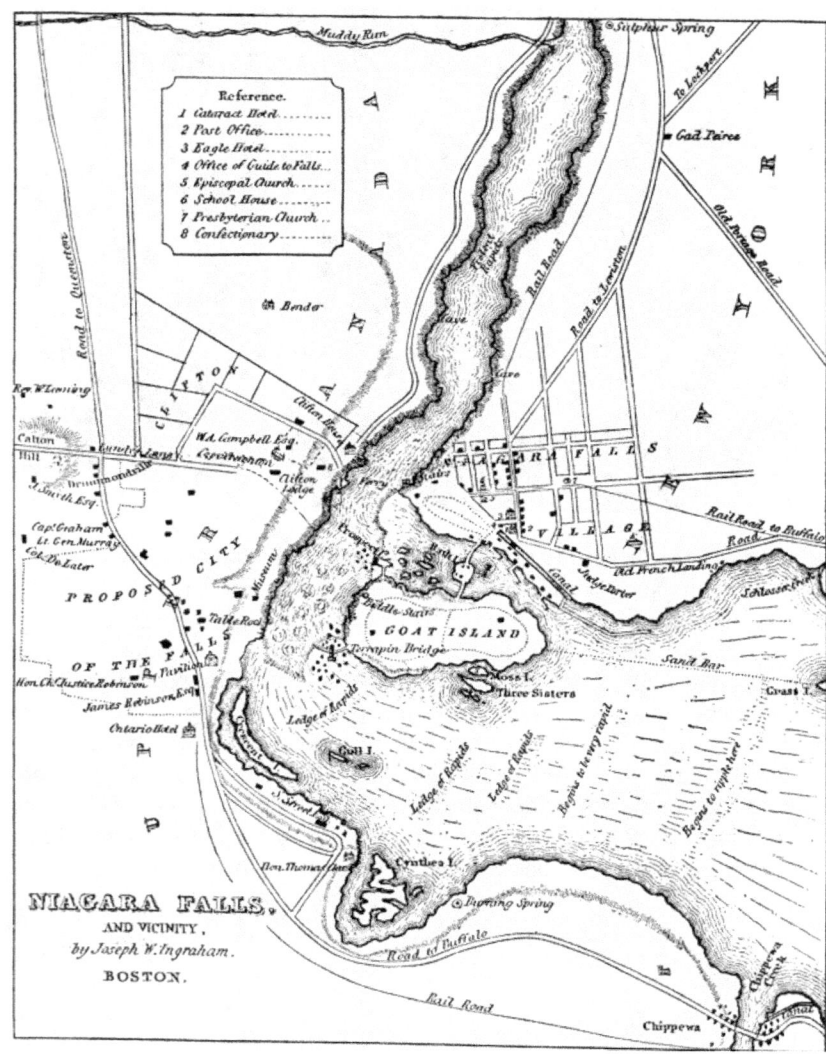

Map of Niagara Falls, 1836 —On the New York side, Niagara Falls is still a "village." The sand bar from Grass Island to Goat Island may have been a means of travel to Goat Island before a bridge was built. Crescent Island on the Ontario side has been filled in as has been the rocky area of the river at Goat Island on the other side of Horseshoe Falls at Terrapin Tower (Adams, 1927).

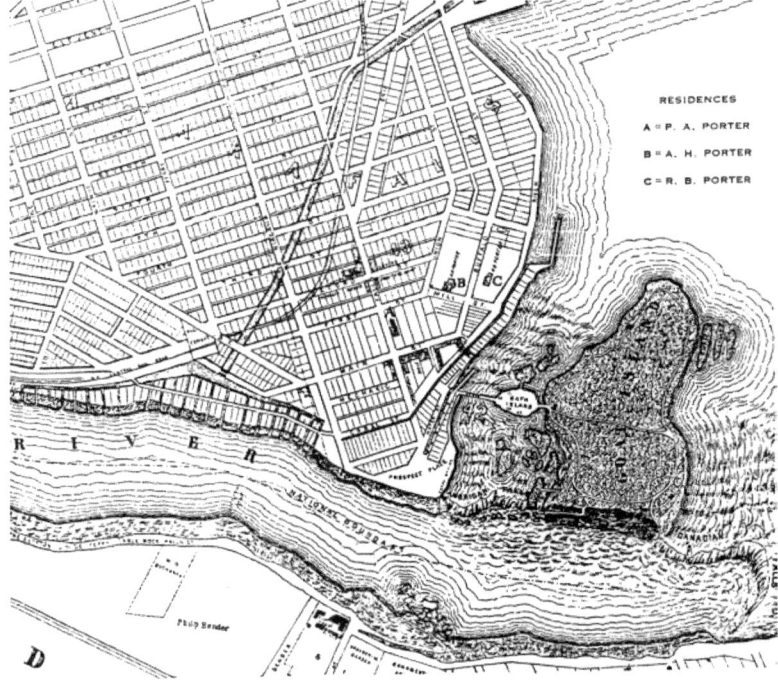

FROM A MAP OF THE VILLAGES OF NIAGARA FALLS AND
NIAGARA CITY, DATED 1856

Villages — From a Map of the Villages of Niagara Falls and Niagara City, 1856 (Adams, 1927). The Porters early developed manufactories and mills. In 1822, Augustus Porter built a flouring mill (later purchased by the Witmer Brothers), and in 1823 a Jesse Symonds did the same "near Goat Island Bridge" (Adams, 1927). Note two shoreline canals built in addition to the later hydraulic canal. The longer one reaching from upriver (upper right) down to the Goat, or Bath, Island bridge was called the Upper Raceway, an extension applied to it in 1826 and utilized by "Ira Cook, William Tuttle, Capin & Swallow," etc. This canal "is supposed to have been built in 1820, when the Porter brothers erected a grist-mill" (Adams, 1927). The smaller shoreline canal was called the Lower Raceway, built around 1845 and on which, "it is said ... a paper-mill, a woolen factory, and a nail-mill were built" (Adams, 1927). These two minor canals and their associated businesses were bought by the State in 1885, and removed. Note the residential district where the Porter brothers resided east of Mill

~ 21 ~

Street, to the present day a small neighborhood of lovely houses. By this time the Three Sisters Islands had not been bridged.

HISTORIC BACKGROUND: THE PORTERS

The nineteenth century Porter brothers, General Peter B. and Judge Augustus Porter, were not simple custodians of an undeveloped park at Goat Island, but were active entrepreneurs. Augustus Porter preceded his brother to western New York when the land was still clothed in virgin timber and the Seneca still held rights to all the land west of the Genesee River. Augustus was primarily a surveyor, marking boundaries for the enormous land purchases of Phelps and Gorham east of the Genesee River, and later for the Holland Land Company, which owned 3.3 million acres of western New York State. Peter B. Porter, a lawyer, had graduated from Yale and was active in the State Republican party at the turn of the eighteenth to nineteenth centuries. In 1800, when all land west of the Genesee River, now divided into eight counties, was organized under the name Ontario County, Peter B. Porter served as its county clerk. Although in many ways a brilliant family, the Porters were to fall heir to a series of disappointments which were to dog them through the early decades of the twentieth century.

Augustus Porter, 1769–1849 — "Pioneer surveyor, 1789. Power Pioneer of Niagara, 1806. First Judge of Niagara County, 1808" (Adams, 1927).

Perhaps their first disappointment was the decision to back Aaron Burr for the scandalous New York State gubernatorial race of 1804, which disrupted the unity of the Republican party by Burr's "disregard of the gubernatorial choice of the regular Republican leadership" (Chazanof, 1970). When Burr lost, Peter Porter lost the clerkship given him by the Republican party in 1797 in return for services rendered to the party "to [punish] him for his political heresy" (Chazanof, 1970). This event highlighted the Porter's willingness to break ranks or take risks, even to the breakdown of the union of the young United States, as many believed Burr was intending, and the division of their political party.

By 1806 Peter B. Porter was elected Congressman by the Republicans, and re-elected in 1810. In Congress, as member of the Foreign Relations Committee, Peter Porter was an advocate of an open war with England both because he "felt bitter over England's conduct on the high seas" and because he "feared her close relationship with the Indians" (Chazanof, 1970), but also because "he wanted to add to his already sizable land holdings and was not averse to expanding into Canada. He sought, too, a monopoly of the shipping trade for his company [Porter, Barton Company] on the Great Lakes. He hoped to take over the business of the wealthy English firm of Robert Hamilton which controlled the forwarding trade on the Canadian side" (Chazanof, 1970). Peter Porter's center of operations was Black Rock, a town just north of Buffalo, and later incorporated into Buffalo.

Once war was declared on June 18, 1812, Peter Porter "promptly resigned his office as congressman to join the actual fighting in western New York and to sell supplies to the army " (Chazanof, 1970). Porter came back to western New York to better handle the business his company would have in supplying American government forces, and to do the best he could to ensure a military success that would enhance his economic interests and the interests of his government, that is, to redress naval offenses to the United States by England, and England's apparent rousing of the Indian peoples against American settlers. He distinguished himself as a soldier in that conflict, and was willing to risk his life and his holdings for a favorable outcome, for Congress granted P. B. Porter a gold medal "in testimony of the high sense entertained by Congress of his gallantry and good conduct in the battles of 1814 ... at Chippewa ... Niagara (Lundy's Lane) [and] ... Erie" (Adams, 1927).

However, the war "seriously interfered with the transportation operations of this firm" to the extent that when their twenty-year lease came up for renewal in 1817, it was "extended for four years in recognition of the forced suspension of the business during the war" (Adams,

1927). The United States Government had purchased the Porter's ships and had them "added to the armed fleets on the lakes" (Porter, 1875). While Peter B. Porter's sphere of influence was international and centered in Black Rock, his brother, Judge Augustus Porter confined himself to what would become the village of Niagara Falls—most of the land of which he, or his interests, already owned (Porter, in Adams, 1927). The war devastated the developing village of Niagara. "At this place nothing was saved, except two or three small dwellings and the log tavern ...," including Porter's home (Porter, 1875). The log tavern was to become the famous Eagle Tavern of Parkhurst Whitney.

     Erie Canal — The Porters also lost their bid to have the course of the Erie Canal run to Lake Ontario at Oswego "and around Niagara Falls to Lake Erie," instead of the present interior route to Buffalo. De Witt Clinton was governor of New York State when both houses of the State legislature voted the survey of lands for the proposed canal routes. Note that this De Witt Clinton was not, as is frequently supposed, the "talented naturalist" of the same name. De Witt Clinton, the naturalist, was "nephew of the seven-time governor of New York State, former mayor of New York City, and a strong political leader in the Senate who became interested in the canal" (Chazanof, 1970). It is perhaps because of the nephew's scientific interest in the canal—in 1810 he was made one of the commissioners to survey and study the canal proposal—that scientific expeditions were launched on the canal after its opening (see section above regarding George W. Clinton).

     Peter B. Porter "favored a canal route that facilitated trade on the Great Lakes, and sought a passage by way of Lake Ontario. [This route] would aid the Great Lakes shipping firm of Porter, Barton and Company at Black Rock, for the canal would run from Albany to Lake Ontario. The Porter, Barton Company had a twenty-year contract with the State of New York that gave the Company a trade monopoly on the Niagara River and over the Portage Road. [The Ontario route would] insure the trade position of Porter's Company" (Chazanof, 1970). Porter was one of the Canal commissioners. The Oswego-Lake Ontario route was not chosen by the Legislature.

     The firm of Porter, Barton and Company came into being with the grant by the state to the partners in the firm of Augustus Porter, Peter B. Porter, Benjamin Barton and Joseph Annin to authorize the company to own the lease of the Niagara portage by an Act of 1803 (Adams, 1927).

Again, Porter wanted the western terminus of the canal to be the Black Rock harbor, rather than the harbor to be established at the City of Buffalo at the junction of Buffalo Creek with Lake Erie. Cargo, destined primarily for Cleveland, was taken by ox or horse teams from Lewiston to Fort Schlosser above the American Falls, and from there poled up the River to warehouses in Black Rock for shipment west. "Starting in 1806, when the legislature had permitted the sale of lands in the vicinity of Black Rock, Augustus and Peter B. Porter had accumulated sizable holdings in that area. They had joined with Benjamin Barton and Barton's uncle, Joseph Annin, in buying some lots along the Niagara River from Fort Niagara to Black Rock and in leasing the landing places at Black Rock and Lewiston" (Chazanof, 1970). Augustus and Peter Porter, in addition to their commercial vessels on the Great Lakes, "supplied the military posts ... at Fort Niagara, Wayne, Chicago, and Michilimackinac .... The firm also handled nearly all the business of the American fur companies as well as that of large Indian traders. With its monopoly of transportation along this much-used route, Porter, Barton and Company controlled the portage business using the Niagara River" (Chazanof, 1970). It was out of Oswego that the Porters received freight from New York to be shipped by them along Lake Ontario, through the portage along the Niagara River to Lake Erie and on to Cleveland and Pittsburgh (Adams, 1927). "Earlier, Peter Porter had tried to influence Governor Tompkins in the selection of a canal route directly to Oswego and Lake Ontario. Clinton claimed that Porter had 'infused his opinions into Tompkins who is profoundly ignorant of the subject, whose opinion [is] not worthy of respect but whose opposition is heretofore indirect.'" (Chazanof, 1970). The Porter brothers also lost this bid, to which they had contributed substantial money and effort.

The Canal was to defeat the Porter's usefulness as conveyors in Lake Ontario and the Niagara portage, since goods could be now shipped directly to Lake Erie from Albany. The opening of the canal had a "paralyzing influence upon the business prospects" of the Porter company.

Government policy — Throughout the career of Peter B. and Augustus Porter, both individuals had striven to manipulate government policy for their own personal economic gain. Although devotion to one's company interests and to one's government do not necessarily conflict, there is some evidence, as given above, that the Porters, especially Peter B. Porter, consistently used political position to further the interests of the family, as when Peter was a member of the Federal Congress, a General of the United States Army, and a commissioner of the Erie Canal.

Chazanof, in his excellent biography of Joseph Ellicott, does not characterize Peter B. Porter's political efforts in this way, yet he presents historical information which, in sum, presents Porter as using his political position as much to his personal gain as for the people he was to represent. Perhaps this attitude was central to the continuous history of failure of the Porter family to succeed in the economic ventures which, although superficially they were in advantageous positions to exploit, yet which eluded their grasp, bringing riches to their rivals. Although they may have been quick to see political and economic opportunity, they either did not integrate the welfare of their company with the welfare of the State of New York or consistently misread developments in the State with regard to western New York. For instance, it was because Joseph Ellicott, Chief-agent for the Holland Land Company, in many ways the Porter's chief political and economic rival in the first decades of the nineteenth century, worked in good faith for the interests of both company and State, that the State favored Ellicott's suggestions regarding the route of the Erie Canal, for instance, over competing claims by the Porter brothers (Chazanof, 1970).

Property at the Falls — As mentioned above, in the section above on land use on Goat Island, Augustus Porter acquired Goat Island in October 1815, sharing ownership of it with Peter Porter after November 16, 1816, when the deed was finally given to the elder brother by the State of New York. Augustus had originally applied for ownership of Goat Island in 1811, but at the time the State needed a place for a second prison—the first having been built in New York City—and was considering using Goat Island for such a purpose, or as an arsenal, especially since western New York was badly prepared for war with England. By 1819, the State had decided to build a prison in Auburn, New York, and the war with England had come and gone. In 1814, Augustus Porter, a judge, "still wanted Goat Island, and he finally outwitted the State, and obtained it" (Porter, 1900) by a legal maneuver.

Although the Porters and other business associates, such as Benjamin Barton and Joseph Annin, Barton's uncle, had purchased the lands east of Prospect Point and upriver to the east of Gill Creek at an auction, February 26, 1805, being acreage on the "Mile Strip," the islands in the Niagara River were still owned by the Seneca, probably as acreage, together with their Reservations, not having been sold to business interests during the Treaty of Big Tree in 1797. The Indian title to all the islands in the Niagara River, previously reserved for them, had been extinguished by the State of New York "only a few weeks before" October, 1815 for "$1,000 cash and $1,500 a year in perpetuity" (Porter, 1900).

In addition to the purchase of river-front property in association with their conveying and forwarding operations, the Porters were "the largest owners of important tracts of real estate favorably located for power development and manufacturing purposes" (Adams, 1927). When the Erie Canal, which opened in 1825, made redundant the Porter's principal sphere of financial operations, they began to turn their attention away from Buffalo and on to Niagara Falls.

"As pioneers in power development the Porter Brothers again devoted their influences and activities to the up building of Niagara as a center of population and commerce. In this new period of their lives the great cataract was to be again the pivot of their public lives. Its hindrance to commerce had been the foundation of their business success; they now saw their opportunity in its vast undeveloped power, which became their hope for the utilization of their large landed estate, as well as for the community in which they dwelt" (Adams, 1927).

On June 24, 1825, within a few years of losing their lease for the portage business along the Niagara River, the Porters issued an "Invitation to Eastern Capitalists and Manufacturers." It is a central document in the "story of Niagara Falls" for it was issued by one of the foremost regional capitalist families of the time and represents the focus and direction of economic development to exploit the falls of Niagara.

As early as 1825, the Porters developed a prospectus that described Goat Island (as Iris Island) to the economic community of the United States as a "situation ... not surpassed, and probably not equaled, in the United States, as a site for the establishment of manufactures," The "adjoining banks appear to have been expressly designed for the convenience of leading water from the river for hydraulic operations. Practically speaking, the extent to which water-power may be here applied is without limit. A thousand mills might be erected with the same ease, and equally accessible, as if on a plain; and each supplied with a never failing water-power" (Porter, in Adams, 1927). The grand canal, or Erie Canal, was only ten miles upriver on a navigable waterway where great transportation lines connect the area with New York City and European markets, and the interior United States along the Great Lakes shipping lanes. Downriver sits Lewiston "the head of the sloop navigation of Lake Ontario and the St. Lawrence" and Canadian and French-Canadian markets and shipping lines out of Montreal. Only ten years earlier, the aboriginal peoples, the Seneca, had owned title to Goat Island.

The environment about the falls is "rich in soil, romantically beautiful in formation, and proverbial for salubrity. The pure and limpid waters of the Niagara" abound. "The extensive forests which border the Niagara, the lake and the canal, and cover the islands in the river, will furnish a cheap and abundant supply of fuel for manufacturing purposes, for many years to come" until transportation lines became established and coal could be imported from the coal mines to the south and west.

Goat Island itself "contains about seventy acres of excellent land, the upper half of which might be covered with machinery, propelled by water-power; and the lower half, situated in the midst of the falls and rapids, where Nature courts the imagination in her most sublime, beautiful and fascinating forms, might be converted into delightful seats for the residence of private gentlemen, or appropriated to hotels and pleasure grounds for the accommodation of the numerous strangers who annually visit this spot" (Porter, in Adams, 1927).

The striking aspect of this document is that in all the subsequent literature cited down to the present day by social forces opposed to spoliation of the very beauties extolled by the Porters in 1825, no reference has ever been made, to this writer's knowledge, of this early scheme to sell Niagara Falls and reduce it to a manufactory. Much of the literature written about the Falls after its establishment as an international park in 1885 has seemed to pick up a kind of literary iconography due to the intense use of rhetorical phrases and artifice. One kind of icon is the list of developments proposed for the "pristine" Goat Island made by people other than the Porters, started by Peter A. Porter in his history of Goat Island (1900) and repeated by park proponents into the twentieth century. This icon is called "proposed uses" after Porter's subtitle in his essay where he lists uses such as "for a sheep pen," State prison, State arsenal, circus ground (P. T. Barnum), picnic ground and terminal of the Erie railroad (Jim Fiske), etc. Cornelius Vanderbilt, Sr. was said to have wanted "to buy it for use as a pleasure ground in connection with his railroads...." Peter B. and Augustus Porter had offered up the Island for sale to capitalists, such as Mr. Vanderbilt, for just such a purpose, although the economic ambitions of the Porter family do not appear in these lists. Porter even seems to condemn such uses of Goat Island by "men seeming to be unable to realize (when they think they see a dollar for themselves)" that Goat Island deserved to retain its beautiful native environment, even though the family had earlier cheerfully offered up the island to capitalists for hydraulic, manufacturing and hotel development.

In the "Invitation to Eastern Capitalists and Manufacturers" is laid out the advantages of exploiting Niagara Falls, which have not changed down to the present day. This document is the first of several to be written during the nineteenth century appealing to capital for development of Niagara Falls. Only details have changed.

Development at the Falls — It is ironic that the very qualities of virginity, tranquility, purity, and natural beauty were to have an appeal to men of wealth who were promised the destruction of the forests to fuel industries since known to be notorious for their atmospheric polluting qualities, especially with the introduction of coal as an energy source. The "salubrious" vicinity of the Falls would become environmentally poisoned in association with the "thousand mills," just as the great manufacturing centers of the industrial revolution in Great Britain became so. Niagara's waters, known for their purity and "limpidity," were guaranteed to become deeply polluted by the "thousand mills," several of them pulp and lumber mills exploiting local forests, to be perched on its banks. Industries would cover the vicinity of the falls, "rich in soil," with manufactories, railroads, storage areas for goods and raw materials and raw dirt roads.

Cheek by jowl could be established hotels and pleasure grounds in the midst of a sea of industrial development. Perhaps in this characterization is a naive quality on the part of the participants generally ignorant of the environmental problems associated with industry, or naively hopeful that with careful planning, such degradation could be avoided. The belief that a garden could exist satisfactorily with machinery operating in it is and was the fond hope of all who sought and seek to develop Niagara Falls, promising to leave its environment intact, or to use the environment as a device to disguise the ugly facets of industry.

Already in 1825 the Niagara riverbank supported: "a large and valuable grist-mill, saw-mill, two woolen cloth factories, two clothier's shops, several carding and spinning machines, a forge, paper-mill, etc." (Porter, in Adams, 1927). Augustus Porter, who resided in the area of Niagara Falls opposite Goat Island, perhaps on the site of the homestead of John Stedman, had erected a large flouring mill, sold later to the Wetmore Brothers. The paper mill, built in 1823 by Jesse Symonds near the Goat Island bridge, was one of the largest in the country (Porter, 1875). In 1826, Porter and Clark built a large paper mill on Bath Island, which was extended later by L. C. Woodruff (Porter, 1875). The famous Cataract House "was built, in part, by David Chapman in 1824" (Porter, 1875). All this development due to the attraction of capital investment to the hydraulic capability of the Niagara riverbanks ended temporarily when the Erie Canal opened in 1825, and such capital attached itself to towns along this important transportation route.

The effect of the opening of the Canal was "to divert all the business of transportation from the old channel, and attract all enterprise and capital, seeking employment, to the numerous villages growing up on the line of the canal. Another injurious effect of the canal on this locality [i.e. Niagara Falls], though beneficial to the new villages, was the large water power it [i.e. the canal] afforded, at points where little or none had previously existed; at Black Rock, Lockport, Medina and other towns west of Rochester, adding greatly to their growth, and proportionally lessening ours" [at Niagara Falls] (Porter, 1875).

Hydraulic canal — The Porters in 1825 appear to have been rich in land if not in capital, and they occupied themselves for the remainder of the century selling off parcels along the river bit by bit, and attempting themselves to capitalize and get developed a hydraulic canal on their properties at Niagara Falls leading from the water in the upper river to the lip of the Niagara gorge just below the cataracts. In 1825 Peter B. and Augustus Porter were willing to subdivide their Goat Island and other Niagara properties, and it is likely that Peter B. Porter tried to officially change the name of Goat Island to "Iris Island" in the treaty with Britain ending the War of 1812, the latter name appearing in the prospectus of 1825, in anticipation of making the land intended for sale to appear more attractive, together with the healthy, clean, loveliness of the pre-sale property. The name "Iris" is in reference to the rainbows generated by sunlight in the mist from the falls.

Judge Augustus Porter did not have sufficient capital to develop a hydraulic canal for it involved "an expense greater than his own means would afford. His heirs, believing in his estimate of the importance of the work, finally succeeded in securing the means necessary for this great public improvement, by a free gift of the water power, and of about seventy acres of land, lying in the village, adjoining the lower end of the canal" (Porter, 1875). This "gift" was part of the conspectus Augustus Porter issued to capitalists and manufacturers in 1847 proposing construction of a hydraulic canal commencing "above the great Falls, and terminating on the high bank about half a mile below" (Porter, in Adams, 1927).

The canal project began when "Caleb S. Woodhull of New York, and Walter Bryant and associates of Boston, in 1852, entered into a contract with the heirs of Augustus Porter" to purchase the properties from the Porters pertinent to establishment of the canal and its operation

(Adams, 1927). About 45 acres of land on the gorge rim were involved, extending one mile down the river from below the American Falls. The Niagara Falls Hydraulic Company was established by the investors. Ground was broken in 1853 and a prospectus was issued by the company in that year.

Industry versus the sublime — In the prospectus, specific reference is made to the possible degradation of the falls area by manufacturing activities. In addition to its advantages for industrial development, "[Niagara Fall's] attractiveness as a watering place will continue undiminished; for the proposed situation of the factories is such as to preclude the possibility of their detracting in the least from the grandeur of the cataract" and the "celebrity which now attaches to the place, as the possessor of the sublimest of nature's works, will not be lessened when it shall be one of the great workshops of the world, sending forth daily the wonderful creations of human industry and skill." The industrialists were acknowledging by this time the "celebrity" attached to the natural phenomenon of Niagara Falls, the association of a wealthy clientele with this "watering place," the sublimity of this "work of nature." It must be remembered that the hydraulic canal was not yet in operation, no tailraces were discharging over the gorge rim, and much of the upper river had not been developed. One must assume that criticism already existed against the industrial development which was already established beside the big hotels on the riverbank above the falls, and such criticism was known to men of capital.

Here, too, is the pitting of the "wonderful creations of human industry and skill" against the "sublimest of nature's works," which will be met with in future arguments for development of the falls, as though if the two were thrown together the value of both would be increased.

As time went by, this company could not meet its financial obligations. The company was sold to a new group of investors in 1856, who called their company the Niagara Falls Water Power Company. This was the "Day Company" after Horace H. Day who owned controlling interest in the company, and after whom the inlet portion of the canal on the upper river was named ("Port Day"). Commercial use of the canal began in 1858, but the full development of the canal was not complete until 1862. Adams (1927) published a print of the unused first water allowed through the canal and over the high bank at the lower end of the canal in 1857.

Frederick Church, painter — It was in the year 1857, too, that Frederick Church first displayed his influential painting "The Great Fall,

Niagara." It was seven feet in length, and three and one half feet in height and depicted the Horseshoe Falls. It "enraptured critics on both sides of the Atlantic" (Roper, 1973) and probably did much to reinforce the international respect for the visual integrity of the falls. Indeed, depictions of the falls abounded for it was "the most frequently described and depicted natural wonder in North America. No other site in the New World was represented so many times by so many artists in so many ways. From 1760 to 1900, countless amateur sketchers and professional painters, foreign and native-born alike, swarmed over the location, picturing the waterfalls from every possible vantage point" (Adamson, 1985). Church's "Niagara" was the climax of American landscape painting of the nineteenth century (Adamson, 1985).

Church's picture became a sensation in New York City where "thousands were crowding into the Broadway gallery daily to see [it]" (Adamson, 1985). When the painting went to England "the English press was ... unanimous in its praise" and contributed greatly to the prestige of American culture and its potential (Adamson, 1985).

The Niagara Falls Water Power Company became financially exhausted in its turn, and its directors, in 1860, sold the company to Horace H. Day outright. Throughout the Civil War years (1861-1865), Horace Day promoted full excavation of the canal.

As the War proceeded, perhaps the reason Day was able to continue excavation to the extent that occurred was due to pressures to generate industrial goods for the Union cause, which promoted expansion of industry throughout the Union-allied states. The war began in a business depression, which may have accounted for the failure of the company established in 1856. Imposition of tariffs baring imports of European goods together with huge government expenditures appears to have created a booming northern economy, at least for company owners if not labor. Contractors to the government made "fabulous fortunes," generating a new class of businessmen. "Particularly ostentatious and offensive were the nouveaux riches and the "shoddy" rich who had waxed fat from war contracts. "Shoddy"—a reclaimed inferior wool substituted by dishonest manufacturers who attained great profits—came to denote any contractor who made exorbitant profits by supplying the government with inferior goods. Most businessmen were honest and conscientious, but even legitimate profits were very high" (Encyclopedia Americana, 1927). Labor was in short supply throughout the north during the Civil War, and this may have crippled completion of Day's hydraulic canal.

By 1877, the canal was one mile long, "with a capacity of about 27,000 horse-power" when, after borrowing and spending "more than

$800,000," including Mr. Day's "entire fortune," "the company had exhausted its resources and credit before the canal could be sufficiently extended to justify lessees in the construction of manufactories" (Adams, 1927).

Schoellkopf — Day's company sold its property at auction, May 1, 1877, to Jacob F. Schoellkopf and associates, of Buffalo, for $71,000, with $5,000 to "settle accounts" bringing the figure to $76,000 (Adams, 1927). Building on the efforts of the past, it did not appear to require vast amounts of capital for Schoellkopf to finally turn the canal into a successful operation.

Jacob Frederick Schoellkopf 1819–1899 —. Founder in 1877 of the Niagara Falls Hydraulic Power and Manufacturing Company (Adams, 1927).

"An old property owner declared when the sale was announced, 'Now we can add a hundred dollars to the price of every lot'" (Adams, 1927). The oldest and most extensive property owners were the Porters. The citizens of the village of Niagara Falls were said to have expressed "much satisfaction" over the purchase by Mr. Schoellkopf and associates, who were considered, at least by Adams, to have "the means to develop the power and the community of Niagara as they conceived it possible and profitable [and] were hailed as a favorable omen of progress and success for Niagara power" (Adams, 1927).

Jacob Schoellkopf had come to Buffalo, New York, in 1843. He was described (Brown & Watson, 1982) as "the young German immigrant who arrived with a few hundred dollars in his pocket and became "King Jacob," founding a family dynasty with almost legendary accomplishments in nearly everything related to capitalism." He opened a small leather shop in Buffalo and, "because he was a hard-working man ... soon had tanneries in Buffalo, Chicago, and Milwaukee" (Brown & Watson, 1982). No doubt the principle source of Schoellkopf's wealth was in supplying leather goods to the Union army—one can imagine the need for leather of the cavalry alone.

"Jacob Schoellkopf may have best exemplified the German immigrant whose diligent enterprise put an end to Yankee exclusivity in [Buffalo's] commercial and industrial power structure" (Brown & Watson, 1982). Yankee may have meant "vintage Anglo-Saxon or 'old family'," according to these authors.

With the success of the hydraulic canal came the fruition of the warning hinted at in the prospectus of 1853. Degradation of Niagara's environment, already stimulating protest before 1853, would soon become critical and would worsen as a necessary consequence of industrial development. The success of the canal would present an almost irresistible momentum for accelerated industrialization of the village of Niagara Falls, the Niagara environment and all of western New York State and adjacent Ontario.

THE VILLAGE OF NIAGARA FALLS

An ephemeral population of visitors, a group distinct from the first set-tlers, soldiers and traders, was of a constant and regular occurrence at the falls, such that James Flint commented on their presence as early as 1822 (Scott and Scott, 1983). These visitors frequently appealed to govern-ment to provide for their ease of passage, such as La Rochefoucault in 1795, visiting Niagara on the Canadian (British) side who "experienced so much difficulty in making a way through thickets, rocks and swamps to points of vantage from which to view the cataract that he was moved to write: "it is much to be regretted that the government of a people which surpass all other nations in fondness for travelling and curiosity should not have provided convenient places for observing this phenome-non at all possible points of view." He considered that for thirty dollars "the greatest curiosity in the known world would be rendered accessible" (in Way, 1946).

While the earliest industries engaged in by the Europeans dealt with the capacity for milling on the river's edge, in particular, by the French, with mills on both sides of the river (Way, 1946), and trade, there were later taverns and other lodgings and related businesses estab-lished based on the presence of spectators, especially after the turn of the nineteenth century and the war of 1812-14 (Way, 1946). Chippawa, in Ontario and just above the falls, possessed two taverns and ten houses. "In 1801, the Reverend David Bacon of Connecticut discovered what was apparently the first public-house at the site of the cataract. He rec-orded: 'There was at the Falls a good tavern where we took breakfast, but there was no other house, and I think there was none on the American side" (Way, 1946, citing Green, no date). One inn, Forsythe's Hotel, in 1819 (perhaps on the Canadian side) "had lately erected a covered stair-way by which visitors could descend into the gorge below the cataract" (Way, 1946, citing Green, no date), a precedent to the Biddle Stairs. The entertainment of visitors as a commercial enterprise increased in the 1820's (Way, 1946). In 1823 the Ontario House and The Pavilion had been built on the Canadian side, neither of which had existed in 1818 and one visitor noted, in 1822, that "the Falls of Niagara are much visited by strangers ... there is a large tavern on each side of the river, and in the album kept at one of these, I observed that upwards of a hundred folio pages had been written with names within five months" (James Flint, quoted in Way, 1946). The Clifton House, Ontario, was erected in 1853.

Tourism appears to have been more fundamental to the settle-ment of the city of Niagara Falls, Ontario, than the twin city on the

American side. A city was to have been created on the Canadian side before 1836. The project "was primarily a financial one dependent on the tourist traffic, it had as a secondary objective, the idea of protecting the Falls from commercial enterprises derogatory to the natural scenery" (Way, 1946). When, however, the American railway systems uniting the settlement on the American side with Buffalo and Lockport were built, "tourist traffic between Buffalo and the Falls was diverted from the Canadian to the American side of the river" and the tourist industry on the Ontario side experienced a setback, such that the project was temporarily forestalled. "The population of the village of Niagara Falls [New York] in 1853 was probably less than 2,000" (Adams, 1927). Some of these people would have worked in the mills established along the upper river bank opposite Goat Island, some in municipal government, such as it was, and some in the hotels built in the area. Although the village sported such a small population, "it had a yearly influx of visitors in numbers up to 60,000 .... Much of this tourism was stimulated by the growth of the railroads .... By 1853 there were six railroads being built or completed in the Falls area" (Scott & Scott, 1983). While the Porters, who essentially owned the aboriginal land on which the village of Niagara Falls developed, kept their eyes on the gigantic prospectus of regional development, local businessmen made their money in the service sector associated with the hotel and tavern businesses. As far as my references indicate, the Porter family never involved itself in the local tourist trade except for their interest in Goat Island. The descendants of Augustus and Peter B. Porter never appeared to have offered up this bit of real estate to development after the 1825 prospectus discussed above, rather concentrating on the disposal of their considerable real estate across the river on the mainland. The possession of Goat Island seemed to give the family prestige, and they embellished it in such a way as to maintain or augment this value. It is perhaps their dilettante interest in history that stimulated the island's maintenance in its primitive state.

Property — With the establishment of the first hydraulic company to develop the canal, in 1852, the Porters disposed of "about 80 acres on the level plain or plateau below the falls, for manufacturing sites, extending about 1 mile on and along the high bank of the river, ... 1100 feet of water front for wharf purposes, above the falls, opposite Grass Island, and ... a strip of land 100 feet wide for the canal, the whole situated within the limits of the village of Niagara Falls" (Adams, 1927). "'All these lands,' it was claimed, 'including their water-privileges and other advantages, together with the exclusive right to construct the proposed canal, were purchased by the company ... and are now absolutely owned by them'" (Adams, 1927). Whether these property rights were sold by the company when it failed, or reverted to the previous owners could not be established here, but the rights most likely stayed with the company and ended be-longing to the Schoellkopf concern, which permanently owned the general outline of these lands up until the State acquired them in the twentieth century.

Grand and internationally famous hotels were erected in the village: those built after the War of 1812 and around the time of the opening of the Erie Canal, as discussed above, and others as the century advanced. The Cataract House was built, in part, by David Chapman in 1824, and has been enlarged from time to time to its present great extent by P. Whitney and Sons. The International Hotel, built by B. F. Childs, and enlarged by J. T. Bush ... ranks with the Cataract House among the largest and best conducted hotels anywhere to be found. The Spencer House and Niagara House here, and the Monteagle Hotel at Suspension Bridge, are all of more recent date, and are all of them hotels of high character and large capacity" (A. Porter, 1875).

In the mid-1940's, for example, the Cataract House "was extended to the river's edge where bath areas were built in the rapids" (Scott & Scott, 1983). "This addition included a ballroom and balconies extending over the edge of the upper rapids .... The unique combination of one of America's most spectacular natural settings and a fine hotel representing the height of mid-19th century luxury often moved visitors standing on the balconies of the Cataract to compose reflective responses to the majesty of the cataract or of life itself" (Scott & Scott, 1983).

"The mood of a visit to Niagara in the 1840's was genteel. Gardens and a fish pond edged the River bank by the Cataract House." In the 1850's the International Hotel was built "across the street from the Cataract next to where the Eagle Tavern had long been established .... The Vedder, also known as the Glove, the Camel, the Frontier House, or the Fall's View Hotel was constructed at Ontario Street and Whirlpool. In 1852 the American Hotel was built .... On Falls Street the Empire House [was] built in 1852 .... These hotels were generally seasonal, but they catered to both short-term visitors and summer guests" (Scott & Scott, 1983).

During the 1850's a suspension bridge was built to Canada and DeVeaux College was established; the town of Suspension Bridge was incorporated into what became the City of Niagara Falls in 1854. The Monteagle Hotel opened in 1855, "and in the following year the very popular International Hotel opposite Prospect Park was expanded. ... Opposite the International, the Niagara Park Place was opened which included reading rooms, billiards, liquor, and a soda fountain" (Scott & Scott, 1983).

Society — It appears that it was during the decade of the Civil War that the peaceful mood within the visitor population and the system erected for its comfort changed for the worse. As discussed in the previous section, the nature of the business community changed perhaps from something that might be described as "old money" to a class of businessmen recently made rich from sale of goods under contract to the Union government—an "ostentatious and offensive" lot. "The prosperity, anguish, and dislocations of war stimulated extravagant living, frantic amusements, vice, and crime .... Entertainment flourished. Theaters featuring comedy were jammed; and parties, balls, and receptions cheered both the indifferent and the afflicted of war. Vice flourished in cities and near army camps, and Lincoln observed that in the wake of war 'Every foul bird comes abroad, and every dirty reptile rises up. These add crime to confusion ... Murders for old grudges, and murders for pelf, proceed under any cloak that will best cover for the occasion" (Hoogenboom, in American Encyclopedia, 1927).

Alfred Porter, one of representatives of "old money" on the Niagara Frontier, characterized the sad state to which society had come in his town: Niagara Falls:

"Few persons in this progressive age will admit that the olden time was better than the new, yet it cannot be denied that the fearful increase of crime, and the sad corruption of morals, now so obvious, both in public and private life, are legitimate fruits of wealth and luxury, and of the overweening greed for money, so characteristic of the present times" (Porter, 1875).

"The War did not stop the flow of tourists to the Falls and Niagara business men continued to build to please crowds" (Scott & Scott, 1983). Competition with tourist businesses on the Canadian side was aggravated with the construction of the Upper Suspension Bridge in 1869 (Scott & Scott, 1983). More bridges, walkways, buildings were added to structures in Prospect Park and in the Goat Island complex. Reports, such as that by Mark Twain in 1869, spoke of "competing hackmen, aggressive salesmen, and Indian craftsmen soliciting tourists in shops, in front of hotels, and even throughout the sloped banks of the Park" (Scott & Scott, 1983).

Economic interests — By the 1870's, economic interests began to overwhelm the city environment near the falls. Residential areas at the river's edge and adjacent to the entrance to the bridge to Goat Island were bought up and commercialized -primarily through the enterprise of a Mr. Tugby, who "dominated the River front to the left of the Goat Island Bridge" (Scott and Scott, 1983). The area of Bridge Street, which connected Main Street to the bridge to Goat Island had been residential before it was commercialized.

The Porter family continued to divest itself of riverfront property to developers, particularly Prospect Point in 1872. Like Goat Island, the Porters had maintained Prospect Point with little development. As part of the agreement of sale of Prospect Point, the Porters demolished the Terrapin Tower, "about the same time that Bridge Street was commercialized" (Scott & Scott, 1983). The "grand old Terrapin Tower was needlessly torn down in 1873 in order that it might not prove an adverse attraction to the interests of a company which had bought and were about to fence in the last spot of land on the American shore, from which a near view of the Falls could be obtained; a point which so long as it remained in the possession of the owners of Goat Island had been free to the world" (Porter, 1900). This last note is an example of the contradictory stance of the Porters. The Prospect Park Company would be one of the last of the private concerns to protest the State's condemnation of their property in the establishment of the State Reservation (Welch, 1903). Fences were built "around much of Prospect Point making the Falls only visible to paying customers" (Scott and Scott, 1983). With Terrapin Tower gone, one of the major prospect opportunities was lost to the people and the attraction to Goat Island was lessened in favor of the business enterprise on Prospect Park—with full cooperation by the Porter owners. The company operating the Park established on this tract "a store, a dwelling, garden rockwork, terraced slopes, flower beds, ornamental trees, fountains, several pavilions, an art gallery, a variety theatre, and machinery for illuminating the Falls" (Scott and Scott, 1983). A fountain was built by the Prospect Park Company in 1877 "of stone, earth and moss in the center of a large fish pond designed in the manner of a modern Disney creation .... Everything was improved for the tourists .... Even an Art Gallery housing paintings, magnifying lenses and stereoscopes was built near the Pavilion ...." Yet the overall impression left in the minds of the visitors seems to be expressed by Henri Rochefort's statement that: 'The banks on the side of the rapids are crowded with pedlars and even fair-stalls. Everything is on sale" (Scott & Scott, 1983). Perhaps as a testament to the impact of the Civil War on the minds of the citizens of Niagara Falls, a monument to Civil War soldiers was erected there in 1876 (Scott & Scott, 1983).

Goat Island — Scott and Scott (1983) report many instances in which it was attested that Goat Island and the American side was one of the most popular attractions in the area, such in the 1830's when "the Canadian side was preferred for its view, but Goat Island continued to attract the most visitors." Again, Goat Island was the most popular attraction in the 1860's, with over 100,000 people predicted in the Niagara Falls Gazette to visit in 1868. "By the 1870's the area [of the falls] had become, to the huge profit and advantage of local business, the greatest tourist attraction in the Americas" (Way, 1946).

The falls received world-wide advertisement in commercial views of Niagara distributed world-wide, which "decorated theater curtains, scenic wallpapers, dinnerware, sheet music, commercial advertising labels, lamp shades, and stock certificates. Niagara Falls also proved to be the single most popular landscape subject among the millions of mass-produced stereographic photographs that flooded the parlors of the Old and New Worlds after 1860. It was truly a theme for Everyman, appealing equally to all classes and levels of taste on both sides of the Atlantic" (Adamson, 1985).

Gradually there came a point where the businesses established to derive economic benefits from serving visitors found themselves in a position to use these visitors in such a way as to maximize the most profits for themselves. They permitted their customers little independence and held the vistas for ransom. High fences were built at Prospect Point and other areas on the shoreline, and visitors were blocked from experiencing Niagara without first paying a fee. Only a few individuals effectively blocked the multitudes from enjoying the great natural vista. A precedent for fencing came in the 1820's on the Canadian prospect when, in a bid to monopolize the tourist business, William Forsythe, owner of The Pavilion Hotel mentioned above and perhaps the earlier Forsythe Hotel (if these were not the same enterprise), "in order to prevent free access to points from which a close view of the cataract could be obtained ... erected fences on the Chain Reserve, a reservation of government land sixty-six feet in width which extended along the top of the Niagara's bank and was originally intended for military purposes" (Way, 1946).

In the 1880's, the village, or City, of Niagara Falls, New York, existed still essentially within a rural regional context. A makeshift, unorganized population of individuals was engaged in an unregulated "tourist industry." The village had a government that was inadequate in legislating and enforcing control over these unsophisticated entrepreneurs, or it showed an unwillingness to do so.

Degradation through tourism — Coincidental with the commercialization of Niagara Falls on both sides of the Niagara River was a distressing spiritual degradation of the Falls environment brought about by the tourist industry. Almost as a form of mockery of the "sublimity" of feeling and inspiration traditionally presented by the unique character of Niagara's environment, the tourist industry presented Niagara's unique character as a freak of nature, not as its highest expression. The fences, through which one could peep at the Falls for a coin, degraded the curious, as though paying for a forbidden glance at a human or animal deformity in a carnival.

As a parody of the awesome grandeur and ominous peril of the high energy liquid landscape, people came to "dare" the natural dangers. As early as 1829, Sam Patch erected 95-foot platforms over the water of the plunge pool beside the American Falls and jumped into the deep waters there to an appreciative audience. In 1829-30 Francis Abbott, characterized as the Hermit of Niagara, presented his conspicuous unhappiness to those who would observe him by suspending himself over the cascading waters of the Horseshoe Falls off the Terrapin Rocks bridge: "From the ends of these timbers he would hang by his hands, his body suspended in mid-air over the abyss, exhibiting absolute fearlessness and strength of will" (Porter, 1900). Stunts were performed by the tightrope walkers Blondin and Willa Hunt in the 1850's and 1860's and people thronged to see these people sent to their destruction as well as to their successful accomplishments. As noted earlier, the blasting of weakened cliff-ledges was announced and people crowded to watch.

"Of all the incidents connected with Niagara, none is more thrilling than the efforts made to rescue Avery from a log in the rapids a short distance above the American Fall, on July 19, 1853. The night before, Avery and a companion had been swept over the river in a boat. Avery landed on a log, but his companion was carried over the Fall. All day long mighty efforts were made to save Avery. Boats, rafts and barrels were let down to him from the Goat Island bridge, and towards evening, just when a rescue appeared certain, the very boat that was designed to carry him to safety struck him full in the breast and knocked him into the river, he was hurled over the Falls to the horror of the assembled thousands" (The Niagara Book, 1901). Hence the name Avery's Rock for one of the rocks in the channel of the American falls. A similar incident happened in 1838 to a Mr. Chapin, who was successfully rescued, after whom Chapin's Island was named (The Niagara Book, 1901).

"Everywhere peddlers hawked their wares, and sideshows erected at every vantage point were filled with freaks assembled from all over the world, so that the vicinity of the cataract had taken on the aspect of a colossal carnival. William Dean Howells, in describing the scene in 1860, mentions the tent enshrining a five-legged calf which was offered as a secondary marvel when visitors were satiated with the Falls. 'I do not say,' Howells whimsically observed, 'that the picture of the calf on the outside of the tent was not a good as some pictures of Niagara I have seen. It was, at least, as much like.' Thus on every hand, there was barbarous incongruity between human debasement and the wonder of the cataract" (Way, 1946).

Way (1946) summed up the crisis developing in the cities of Niagara Falls: "Since it is almost a truism that uncontrolled enterprise in any sphere is apt to mistake liberty for license, it is not surprising that the Falls had now [in the 1870's] been for at least a generation the scene of unprincipled exploitation of the tourists by rapacious cabmen and others practiced in the art of polite robbery .... Travelers were regarded as lawful prey, fees being exacted for every service and no service at all. By 1870 the annoyance and humiliation to which tourists were subjected on both sides of the cataract was fast approaching a climax. Already many persons had concluded that a change was imperative and were expressing their conviction that the Governments of Canada and the United States should assume responsibility through the establishment of national parks."

Effects of the Civil War — The Civil War in the United States generated a substantial loss of human life, with more men dying from bacterial causes associated with disease and primitive medical care than in the battles themselves. This situation duplicated the situation for Britain earlier in the Crimea, where "less than five thousand soldiers had succumbed to enemy action, more than sixteen thousand to disease" (Roper, 1973), a condition to which the British Sanitary Commission responded with a vengeance. During the war "humanitarianism flourished and intellectual life was not neglected. The United States Sanitary Commission and the Christian Commission of the Young Men's Christian Association not only helped care for the wounded but also improved the morale of able-bodied soldiers" (Hoogenboom, in Encyclopedia Americana, 1927). Frederick Law Olmsted was elected executive secretary of the United States Sanitary Commission in 1861, a Federal commission established by Lincoln in the face of an inept and hostile Medical Department of the United States Army.

Locally in Buffalo, in association with the war and its ending, socially conscious activities included the Ladies General Aid Society who made shirts, drawers, socks and bandages for the Union army, administered the Soldier's Rest Home and "several Christian benevolent groups catered to the spiritual needs of wounded local soldiers. While the female members ... engaged in these and kindred activities, the men ... donated some of their ... wealth for the foundation of the Buffalo Historical Society, the Fine Arts Academy, and the [Buffalo Society of Natural Sciences]" (Goldman, 1983). The Buffalo chapter of the Young Men's Christian Association shared its facilities with the young Society of Natural Sciences.

But above all, the Civil War resulted in the triumph of a centralized federal government over state's rights—the rights of the nation over those of the states. It gave, then, impetus to the development of the bureaucracies that would govern the natural resources of the nation, for instance the United Stated Forest Service under Gifford Pinchott, friend of Olmsted, appointed forester to the Vanderbilt Biltmore Estate which Olmsted designed and whose "large-scale example of practical forest management," the first in the United States, was conceptualized by the great designer, and implemented by Pinchott. Ultimately would come the National Park Service. A strong, new superior government would become the source of higher authority for problems not handled very satisfactorily by state and municipal governments regarding issues that transcended national and state boundaries.

"During and after the Civil War, the United Stated developed a sharpened consciousness of their unity as a nation. ... Undertakings of national scope came into being, and certain ideas spread, as though by contagion, nationally. National banks were superseding state banks; national securities were a favorite investment. ... A national Academy of Sciences was founded, a National Commissioner of Agriculture was appointed, and a National Department of Education established. For the promotion of industrial and agricultural education on a national scale, Congress appropriated a vast domain from the public lands ... The national park idea was anticipated in the reservation of the Yosemite Valley, and the idea of rural parks for cities was taking hold across the country from New York to San Francisco" (Roper, 1973).

On January 9, 1879, Governor Robinson of New York State submitted a message to the New York State Legislature requesting the appointment of a commission, in conjunction with a similarly appointed commission by the Province of Ontario, to develop a plan to "remedy ... abuses" to which visitors were presently subjected at Niagara Falls.

OLMSTED, YOSEMITE AND THE NIAGARA RESERVATION

Although many people contributed to the complex process of establishment of the Reservation at Niagara, the interest of the distinguished American landscape architect, Frederick Law Olmsted, was the keystone, the vital interest. He was the nucleus round which efforts to protect the cataracts of Niagara from development by private interests were concentrated. After the Niagara Reservation was established, it was his and his partner, Calvert Vaux's, conception of appropriate action toward improvement of this property that has been the source of guidance through a century of park administration.

Frederick Law Olmsted — Landscape architect. National Park Service, PBS.

Olmsted was and is primarily known for creating pastoral landscapes in city environments and articulating the doctrine of the public park. His celebrated designed parks, beginning with Central Park in New York City, enriched the urban lives of people living in many of the larger cities in the United States. This enrichment was deliberate and inherent in Olmsted's concept of urban reform, the democratization of the ownership or stewardship of natural areas by government.

Perhaps it was because Olmsted was the best, or most successful, craftsman of his time in the art of landscape architecture in this country that he could appreciate the value of the most breath-taking scenery of North America—the outcome of natural processes. Olmsted, who could weigh the practicalities of design implementation and design effectiveness in the artificial environments of cities, could see the irreplaceable, unattainable grandeur of natural design inherent in the native ecosystems and physiography of the great aboriginal landscapes of the United States. Olmsted's urban parks were the antithesis of or the balance with the urban industrial environment, just as natural landscapes were of the city landscape. They were necessary to a balanced set of opportunity for the well-being of society.

The Movement — Olmsted's opportunity to participate in the movement to make the preservation of North America's best natural landscapes a government concern came after President Abraham Lincoln, in 1864 during the Civil War, signed into law the first designated state park in the United States. The proposal was submitted to Lincoln by California Senator John Conness with respect to the Mariposa Big Tree Grove and Yosemite Valley. "The United States national parks were born at that moment although Yosemite was not yet national, but state-owned" (Todd, 1982). Olmsted was appointed as one of the commissioners authorized by Governor Low of California to develop policy regarding this land. This would be the first time such a policy would have been formulated. Olmsted's fellow commissioners elected him their chairman.

Inspired by his new authority, Olmsted "had the territory surveyed and mapped in order that roads to and through it could be planned according to his detailed instructions." This was accomplished by Clarence King and James Gardner, employees of the California State Geological Survey", authorized by one of the commissioners, J. D. Whitney (Roper, 1952). Olmsted personally advanced money to have this done (Todd, 1982) prior to the State's designation of funding. The next thing Olmsted did was to write the preliminary report on the management of Yosemite, the authorship of which "logically fell to Olmsted," (Todd, 1982), perhaps because he was "the man best qualified on the commission, and probably in the country, for this unprecedented task" (Roper, 1952), presumably because of his earlier association with the design and purpose of Central Park.

Roper (1952) declared, in the publishing of this document in 1952, using italics for emphasis, that with "this single report, in short, Olmsted formulated a philosophic base for the creation of state and national parks." It is a very interesting document and did indeed embody philosophic principles; it did attempt to establish a land use policy for areas of scenic grandeur for the public good. In this document, Olmsted had "demonstrated the qualities of a conservative social reformer with a distinctive theory about the role that public parks ought to play in a democratic society." The report was "the first systematic exposition in America of the individual's right to enjoy large, impressive public reservations of natural scenery, and, also, the government's obligation to protect him in the exercise of that right" (Todd, 1982).

The present writer has found some difficulty in finding justification for giving this document the momentous historic influence attributed to it, and the "landmark" role the document is suggested to have played in the continuation of the movement to preserve significant natural areas in the United States. As far as I can see, in the two publications examined with respect to the Yosemite report (Roper, 1952; Todd, 1982) the only exposure this document ever had was private: to the initial Yosemite Park commissioners who later were supposed to have suppressed it, and several visitors, including the editors of two important eastern newspapers, the lieutenant governor of Illinois, the attorney general of Massachusetts, the speaker of the U.S. House of Representatives, to whom the report was presented around a campfire in Yosemite Valley. Then it disappeared for nearly a century (Todd, 1982). Perhaps it would have been a landmark if the document had been made public, and if other documents and legislation, could be demonstrated to have evolved from it. It is possible that such may not have been the case and the important concepts forming the content of the Yosemite report were more powerfully disseminated in Olmsted's inspiring conversations with influential colleagues.

Todd (1982) suggested this was the case when he indicated that when Olmsted was actively working to establish the Niagara Reservation by developing a proposal and petition to impress the Governor of New York State, Olmsted made no use of the Yosemite report, which "had had no real impact ... because of its suppression ...." It is odd that, although Olmsted appeared to have the proper audience later at Niagara before which to reissue the substance of the Yosemite proposal, he appeared to choose not to do so.

Not only did Olmsted's report never become generally circulated where it could do the greatest good, but he himself returned East to continue his career while still chairman of the commission of Yosemite. Without strong leadership, Yosemite began to disintegrate as a preserve of natural scenery. No one, commissioner or otherwise, fought to give Yosemite the government funds required to operate (Todd, 1982). Olmsted, back in the East and accepting architectural commissions, knew how his report was being suppressed and that Yosemite was being compromised, but rather "than involve ... [himself] in a public quarrel ... he chose to let the alleged injustice pass and paid no public attention to it. Governor Low officially acknowledged [Olmsted's] resignation from the Yosemite Commission in 1867" (Todd, 1982).

Yosemite — Yosemite Valley, 1865, a painting by Thomas Hill. This is Yosemite as Olmsted would have seen it. Wiki Commons, New York Historical Society.

It seems less ironic, as Todd suggested, that Olmsted's report disappeared than that Olmsted abandoned the first state park, of the commission of which he was chairman, in order to pursue his own career. Private exploitation, so well displayed in the evolution of Central Park and other cultural achievements in New York City, and which continually plague government-sponsored preserves, increased in the preserve at Yosemite after Olmsted abandoned its leadership. According to Todd, 1982, it was the state of California that "allowed it to fall into neglect." It was the naturalist John Muir who "launched a relentless campaign to save Yosemite from logging, grazing, and many other kinds of misuse," (Todd, 1982). Muir, according to a recent account, "became the preeminent champion of wilderness, at a time when preservation was near heresy" and that he "was a consort of presidents and founder of the

Sierra Club ... his tireless lobbying was crucial to the establishment of the national park system" (Book, et al., 1988). If only Olmsted had pursued his leadership at Yosemite, and Muir had had such a plan as Olmsted's in his own arsenal, the Yosemite outcome might have been much different. In 1906, however, the State of California relinquished its title to Yosemite and it became Yosemite National Park. Yosemite Park was a very important precedent and probably can lay claim to being both the first state park (1864), but the first national park as well, as long as one does not quibble about legislation. The federal government had made Yosemite a state park when it had no program of its own. Yellowstone, the first legally unambiguous national park was established later, in 1872. Yosemite was to continue to be the sacrificial lamb of the national lands movement when the Hetch Hetchy Act permitted damming of the river to supply additional water to San Francisco in 1913 during the Woodrow Wilson administration (Albright, 1985), and in 1917 when it was proposed to permit fifty thousand sheep to graze there. The Sierra Club had continued and still continues to block these decisions to destroy native ecosystems in national lands.

Calvert Downing Vaux — Worked with Olmsted in the 1850's to design Central Park, then again on special projects including the Niagara Reservation (Natl. Assoc. for Olmsted Parks).

Olmsted and Niagara — Niagara Falls, New York, was much closer to Olmsted's center of operations, occurring in the state in which he, together with his partner, Calvert Downing Vaux, had established his reputation. Niagara was closer to where most of his influence was concentrated, and where his powerful friends in the eastern establishment were located. Still thinking of Yosemite, it was in the east where "Olmsted worked, frequently behind the scenes, to obtain signatures from influential friends for a petition to Congress on behalf of the sanctity of Yosemite" (Todd, 1982). Niagara became Olmsted's "second great opportunity to champion the cause of conservation." It would be a chance "to save the natural scenery around Niagara Falls from the commercialization and blight into which it had fallen," (Todd, 1982).

Niagara Falls had by around 1865 become "the most popular tourist attraction in the United States" (Todd, 1982). The tourists were a large unorganized constituency; many of the most influential and articulate had been victimized and exploited by those who made money from their intense interest in seeing and experiencing one of nature's greatest phenomena. Visitors who wrote of their experiences deplored the abuses there and many suggested remedies. The painter Frederick Church, for example, "warned of the rapidly approaching ruin of the scenery," (Todd, 1982). It was a situation only looking for leadership.

Since one of the new commissions Olmsted had received upon return to the east coast was the design of several urban parks in the city of Buffalo, New York, Olmsted was in a position to observe first-hand the urgency of the situation at the Falls, some twenty miles north of Buffalo. Olmsted met with his friends the architect H. H. Richardson and William Dorsheimer, both also with professional interests in Buffalo, "to discuss what might be done to combat the desecration" (Todd, 1982). Richardson and Dorsheimer were both Harvard graduates. Richardson was highly regarded then and was to become known as the greatest architect in America in the nineteenth century (Fox, 1986). Dorsheimer was an important factor in this meeting. He was a prominent Buffalo lawyer, and was to be twice lieutenant-governor of New York State, and an advisor to Governor Grover Cleveland at the critical time when legislation was to be passed to set the land at Niagara into the public domain. Dorsheimer was later appointed one of the first Commissioners of the State Reservation, and its first President and was instrumental in promoting Richardson and Olmsted's careers in Buffalo and the State, and both city and state are richer for that patronage.

Perhaps it is indicative of the general mobility of professionals in the United States in the nineteenth century that James Gardner (later spelled Gardiner), who assisted with the surveying of Yosemite under Olmsted and Whitney, the latter a commissioner of that park and Gardner's superior in the California State Geological Survey, was now director of the New York State Survey. Gardner had studied at the Yale Scientific School (Roper, 1952). Ten years after the Olmsted-Dorsheimer-Richardson meeting at Niagara Falls, Governor Robinson of New York appointed the commission to create a plan for a reservation at the Falls, with the collaboration of representatives from Ontario. Gardner completed the survey of the American side which accompanied the preliminary report for the year 1879, published in 1880 (Gardner, 1880).

Free Niagara — Olmsted organized the famous petition for establishment of the Reservation, with the help of friends Dorsheimer, Frederick Church and Charles Eliot Norton, and supported by 700 signators: "some of the most distinguished members of the Anglo-American community who supported the preservation of Niagara" (Todd, 1982; for a partial list and an excellent history of the movement to preserve Niagara and the significance of its administration for the first twenty years of its existence see 19 Ann Rep Comm, 1903). Alonzo B. Cornell, the Governor succeeding Robinson, made no move to establish Niagara Falls in the public domain in spite of these activities.

Blocked politically, recourse to educating the public on the issues and involving them in the debate was made. Olmsted and Norton employed Henry Norman, a Harvard graduate, and Jonathan B. Harrison, who, previous to being a journalist, was a Unitarian minister (Todd, 1982), to provide copy for newspapers in Buffalo, New York City and Boston regarding the Niagara theme. Harrison later became secretary of the Niagara Falls Association, formed in New York City in 1883, and he was responsible for "flooding New York State with letters, pamphlets, and petitions" with respect to creation of a State Reservation at the Falls (Todd, 1982). Norman, for example, wrote a series of letters to the New York Tribune, the New York Herald, the New York Evening Post and the Boston Daily Advertiser (19 Ann Rep Comm, 1903).

Olmsted had many New England connections at Boston and Harvard University, where Dr. Asa Gray was developing the botany department. Olmsted was probably the one who captured the useful botanical testimonial from Gray's good colleague, Dr. Joseph Hooker of Kew Gardens, both distinguished plant geographers, regarding the scientific value of Goat Island with its extraordinary plant diversity. This testimonial was paraphrased by Olmsted in the Gardner survey of 1880 (see section on Hooker), and has been paraphrased by Reservation advocates to this day.

In the creation of the Reservation at Niagara, Olmsted showed his ability and commitment, skillfully using his social theories, his friendships and ability to inspire as the cement holding together a group of colleagues and professionals with a mutual interest in preserving Niagara. Olmsted was the force behind the momentum of their progress. His theories were the plausible hypotheses with which to move ahead in the great experiment of a protected Niagara. They were theories only, but the anticipated results were of too great a value to ignore.

Grover Cleveland was next to succeed to the governorship of New York, and perhaps it is an example of his presidential caliber that it was his administration that established in law the State Park at Niagara, now the oldest State Park in the United States.

"At the time, no tradition of great scenic parks existed anywhere in the world: to protect an area and conserve it for recreational enjoyment was a policy that had never before been adopted for the management of land from the public domain. Surprisingly, there was no strong organized public movement in favor of such parks, and Congress did not seem to have any special interest in the idea," (Todd, 1982). Niagara was to provide an example of "organized public movement in favor of such parks." The federal government recognized Olmsted's contribution in using his achievement as the basis for awarding the Reservation its placement in the United States Registry of National Places (Greenwood, 1976; Fox, 1986). For an excellent review of the steps taken to preserve Niagara see 19 Ann Rep Comm, 1903.

The establishment of Yellowstone in 1872 was the first instance of "the principle of governmental authority to protect and preserve extraordinary phenomena in natural scenery" (19 Ann Rep Comm, 1903) (rhetorically speaking, for Yosemite had that distinction). In 1899 New York State Governor Roosevelt visited the Reservation (16 Ann Rep Comm, 1900). At first reference was made to "the indifference of the State of New York to this first far note of a coming doctrine" (19 Ann Rep Com, 1903). The campaign to protect Niagara had required lots of public addresses, newspaper and magazine articles, as such a concept had not been thought of before: "the campaign of education conducted by the advocates of the Reservation ... have done a great deal toward forming a more intelligent public opinion on this subject, and have materially advanced the movement for the protection of American scenery throughout the country" (19 Ann Rep Comm, 1903).

The Niagara Reservation was also an experiment in the ability of state government to develop departments or offices to oversee the protection of additional significant natural areas in New York State, and their regulation, and its integration into other critical areas of state stewardship—the management of the natural richness that is the heritage of each of the fifty states in the Union. "There are in this State not a few such places and objects that might with general approval be taken by the State and made public possessions for all time. Money expended in fostering a love for natural scenery and stimulating a popular interest in national or local history must be regarded as money well invested" (15 Ann Rep Comm, 1899). The eyes of the Nation, and possibly also of Canada, would be on New York State, which had legislatively devised much enlightened state legislation in many issues throughout its history. The problems that would beset this Reservation would be models for problems and their solution in future policy in the development of public lands.

It was this legacy of Frederick Olmsted, his colleagues and the Niagara Association, an organization he helped found, that prompted the Commissioners to protest later, when the integrity of the Reservation at Niagara was under attack by local industrial interests competing for the natural resources the Reservation was to protect that "it should not be forgotten that the Reservation really belongs to the State, to the whole State and not to any portion or section of it. The organized movement for the protection of the scenery of the Falls had its origin in the great city at the mouth of the Hudson. The same city is assessed for more than one half of the State taxes. Local interest in the Reservation is entirely subordinate to the interest of the State" (15 Ann Rep Comm, 1899).

The Commissioners of the Niagara Reservation — The first Commissioners of the new State Reservation at Niagara appointed by Cleveland were William Dorsheimer, M. B. Anderson, J. Hampden Robb, Sherman S. Rogers and Andrew Haswell Green (for list of Commissioners, Presidents, Secretaries and Treasurers, their home-towns and terms of office till 1903 see 19 Ann Rep Comm, 1903).

Thomas V. Welch was appointed first Superintendent of the Niagara Reservation. Mr. Welch was responsible for overseeing the implementation of policy in the new State Park. His dedication and skill are detailed in the series of annual reports made by the Commissioners appointed to regulate the new Reservation, and printed by and for the State Legislature. Each report had a section written by the Superintendent.

Perhaps it is because the Commissioners asked David Day, a botanist with the Buffalo Society of Natural Sciences. to conduct a floristic survey of the new Reservation and adjacent lands at Niagara Falls that lead Welch to donate a series of his copies of the annual reports to the research library of that institution, which was to become the Buffalo Museum of Science. It is Welch's copies, signed "Compliments of Thomas V. Welch", which were consulted for this paper.

Mr. Welch served "upon many of the most prominent committees of the [New York State] Assembly and took an active and prominent part in a great deal of important legislative business, but the legislation to which he devoted himself heart and soul and into which he threw himself with an all-absorbing energy and purpose, and that with which his name will ever be inseparably associated, was that which had to do with the creation of the New York State Reservation...if it had not been for the work of Mr. Welch the measure would not then have become law..." (20 Ann Rep Comm, 1904). Mr. Welch had been clerk of the village of Niagara Falls, a member of the board of supervisors, then chairman of the board, then member of the New York State Assembly. He served as Superintendent from 1885 to the year of his death in 1903.

THE REPORT OF THE NEW YORK STATE SURVEY (1880)

In Governor Robinson's message to the legislature in January, 1879 (Robinson, in Report of the Executive Committee of the Niagara Falls Association, 1885), specific reference was made to the protection of visitors, and, although New York and Ontario had civil jurisdiction over the Falls of Niagara, yet "in one sense, the sublime exhibition of natural power there witnessed is the property of the whole world." That there was an agenda already in place is referred to by the sentence: "There can be no doubt that many persons abstain from visiting the Falls in consequence of the annoyances referred to, nor can there be any reasonable doubt that the removal of these objections would largely increase the number of visitors annually." As already discussed in the previous section, there was competition between the Canadian and American cities by the cataracts for the money visitors brought with them in their desire to see the falls. There was extensive building and expansion of commercial enterprises and the introduction of rival spectacles and entertainments to lure increasing crowds to the tourist infrastructure in place in the prospect areas of both countries. In one sense, Robinson's message was not only an appeal to protect the ephemeral visitor population at the Falls, but an appeal to devise a plan to "increase the number of visitors annually." It appears that Robinson's message was protective of the tourist industry in place at Niagara—and a plan should be devised for its intelligent regulation. No reference is made to industrial development, except perhaps for the criticism that the "most favorable points of observation around the Falls are appropriated for purposes of private profit," this as distinct from the shores which "swarm with sharpers, hucksters and peddlers, who perpetually harass all visitors." No reference is made either to a deteriorating environment and its restoration.

However, there are two published versions of Robinson's 1879 message to the Legislature. The one quoted above by the Niagara Falls Association, to be discussed below, and one, presumably the official one, as it is a Senate document, contained in the first annual report of the Commissioners of the Niagara Reservation published in 1885. In this second version of the message, the remarks of Lord Dufferin, Governor-General of Canada, with whom Robinson collaborated on introduction of the Niagara issue to their respective legislatures, are included. Robinson acknowledged that it was actually Dufferin's idea for an international park, "inclosing a suitable space on each side of the river from which all the annoyances and vexations referred to should be excluded." Furthermore, in this park there should be "no attempt at landscape ornamenting in the vain hope of adding anything to the natural attractions of the falls," and that such areas are "to be kept sacred to the free use" of international visitors. Dufferin could not envision why such a plan would cost much—it would simply be an undeveloped area, requiring little capital investment by the respective governments, "but with a mutual understanding as to the general regulations to be enforced on either side." It is interesting that Dufferin's reference to the environment was deleted from the executive report of the Niagara Association, as was the fact that it was Dufferin who made the suggestion to Robinson. Dufferin's proposal that there be an international park with mutually agreed upon rules of operation was also deleted. The reason for the deletion by the Niagara Association appears to have been because the men who came together in New York City and formed the Niagara Association to promote legislation to protect the falls in the New York Legislature "decided to give up the international park idea because of the difficulty of meshing New York's moves with Canada's and to appeal instead for public backing of a plan to buy the land around the falls for a state reservation" (Roper, 1973, according to correspondence between Olmsted and Charles Elliot Norton in 1882).

The Commissioners of the New York State Survey

The Legislature did not hesitate to respond "by a joint resolution of the Legislature of that year" (1879) to direct the Commissioners of the New York State Survey "to inquire, consider and report what, if any, measures it may be expedient for the State to adopt for carrying out the suggestions contained in the annual message of the Governor with respect to Niagara Falls" (Report of the Executive Committee of the Niagara Falls Association, 1885). The Commissioners were W. A. Wheeler, Robert S. Hale, William Dorsheimer, Francis A. Stout, George Geddes (son of James Geddes, engineer of the Erie Canal—the son being a specialist in law, engineering and farming,) and F.A.P. Barnard, President of Columbia College (Roper, 1973).

Director of the New York State Survey at the time was James T. Gardner, who was directed "to make an examination of the premises, and prepare for their consideration such a project ... and they associated with him Mr. Frederick Law Olmsted" (Rep. of the Exec. Comm, Nia. Falls Assn., 1885). Note that the spelling of Mr. Gardner's name here follows that used in the final report submitted to the Legislature in 1880—it is sometimes spelled Gardiner or Gardener. The resolution laid out an agenda that was a striking contrast to that presumably proposed by Robinson, according to the Niagara Association. The response of the New

York State Legislature promoted an environmental, not a tourist, agenda, that is, the entire point of all interest was the integrity of that area the tourists were visiting. The intent of the legislation was not to provide for tourists, but to protect the object of their interest and the freedom to enjoy that object without distraction.

The State Legislature appointed the Commissioners and Gardner and Olmsted to determine "how far the private holding of land about Niagara Falls has worked to public disadvantage through defacements of the scenery; to determine the character of such defacements; to estimate the tendency to greater injury, and, lastly to consider whether the proposed action by the State is necessary to arrest the process of destruction and restore to the scenery its original character" (Report of the Executive Committee of the Niagara Falls Association, 1885). Note that "scenery" here is essentially synonymous with "environment" and is botanical in character, as will be seen below.

In that same year, the Commissioners, Gardner and Olmsted came to the area and contributed their findings, which, together with the determinations of the Commissioners, were published in 1880 in the Report of the New York State Survey for the year 1879 entitled "Special Report of New York State survey on the Preservation of the Scenery of Niagara Falls, and Fourth Annual Report on the Triangulation of the State for the year 1879." The special report is 42 pages long. In it Governor Robinson's full message to the Legislature is included, together with Lord Dufferin's comments.

The State Commissioner's Report

The plan for the preservation of the scenery at Niagara Falls is composed of (1) the report of the Commissioners of the New York State Survey, (2) the report of Mr. Gardner, and (3) that of Mr. Olmsted. There were, in addition, two illustrations, a series of heliotype prints, the signatories of a petition in favor of the establishment of the reservation, Governor Robinson's 1879 message, and a facsimile reprint of Hennepin's description of the falls at Niagara.

The First Illustration — To give it the most dramatic thrust, the first product of the study is an illustration of the object of the plan, the product of the study itself. It is a fold-out illustration drawn by Francis Lathrop and engraved by a Mr. Marsh entitled "Ideal view of the American Rapids, after the Village Shore and Bath Island are restored, according to the proposed plan." Such a picture was to be shown in contrast with the series of heliotype prints showing the disfigurements to be remedied—the industrial presence on the river banks just above the American Falls and on one of the developed islands in the Goat Island complex—Bath Island.

Ideal view — This is Lathrop and Marsh's illustration of the Ideal View of the American Rapid, after the village shore and Bath Island are restored. The bridge to Bath Island is in the upper middle. The islands in the river and the American shore are depicted as well forested.

Presumably this illustration would not have been included or given such prominence had it not been an important part of the intentions of the Commissioners, Mr. Gardner and Mr. Olmsted.

In the drawing, the sight of the City of Niagara Falls is completely masked from view excepting perhaps four church steeples, a possible factory smoke-stack and a tall neo-classical building. These, of course, given that the view drawn is at considerable elevation to an observer on the ground, would be not seen by visitors to the area in question. The shorelines support unbroken masses of vegetation. The mainland shows a closed canopy of deciduous trees giving way to shrubs, smaller trees as perhaps native willows may appear, and Arbor Vitae or Juniper trees on the river margin. The model for this scenery was the intact vegetation of Goat Island.

The banks are not riprapped and show irregular erosional effects, including the slumping slope with a many-stemmed tree group inclining into the river, much as can be seen today on the wet, sagging slopes of the north shore of Goat Island. The trees are of considerable age and show no attempt at pruning: broken major branches are conspicuous on some trees, stems distorted by natural processes are shown, a tall old tree bole with its canopy blasted away by natural events is shown on what is probably Luna Island in the lower right. No attempt at silviculture is evident on the character of the restored forest. Dead trees are an important component of the natural drama. Bath Island, the island upriver shown as accessed by the bridges of the time, is illustrated with its downriver acreage significantly reduced—all its "made land" has been washed away and the island's supposed natural boundaries are restored. Also, the original, primeval arboreal character of the islands and river margins is depicted as dense with coniferous tree species.

Note that in the heliotype print constituting Plate III showing the aspect of Prospect Park in 1879 as seen as looking north across the American Falls from Goat Island, and that the Park is still extensively wooded for all the buildings and other improvements reported to have been constructed on it. The city is completely masked from view by the tree canopy. Gardner, in his accompanying report, praises the cluster of trees there, but deplores the fact that shrub and herbaceous layers, being "rich masses of woodbine," were destroyed in the building of walls and structures.

To resume, there are no roads or paths or structures of any kind shown in the illustration, nor are any human beings shown at all. Note that the shoreline of the mainland does not depict the limestone flats that probably occurred there at the time of the drawing, but perhaps were built over or appeared to the illustrator to represent human degradation of a natural landscape feature but which were, in fact, created by the dynamics of the river itself. Note that the caption said the depiction is an ideal situation. To give the viewer today some idea what the view up the American rapids would look like after restoration, one may stand on one of the two farther bridges on the Three Sisters Islands where the vegetation comes uninterrupted to the river bank, and where most of the developed shoreline in the Canadian and on Goat Island is blocked from view.

The helioprint of the Goat Island shore (Plate IX) gives a poor impression of their grandeur, hence the drawing. In the list of illustrations and maps at the beginning of the report, there is a caption reading "Plates showing the part of the banks of the American Rapids which still remain in natural condition," but it appears that these plates were not included in the final edition printed for distribution in the Legislature. Although the helioprint "photographs" were apparently excellent for views showing degradation of the shoreline, their artistry was completely inadequate to represent the value of the forest beauty—these had to be drawn to be adequately represented.

It is apparent from the illustration that the primeval vegetation exhibited on Goat Island was to be used as a model for the riverbanks and other areas interior to the river. It is within this context that the "improvements" to be suggested by Olmsted and Vaux years later were to be placed and within which tourists would find what they had come miles to find. This was the object of State involvement. Note that, outside of the bridges, there seems to be little requirement for State revenue to maintain such an environment, or landscape. No pruning, no mowing, no interference. This was in keeping with the plans to return the area to its primitive aspect.

The Report of the Commissioners of the State Survey

The Commissioner's topics proceeded as follows:

1.

The duty of the Commissioners was defined as investigation into the defacement of scenery, degradation of landscape, necessity of government action to arrest further deterioration. No reference was made to the duty of the Commissioners to investigate conditions with respect to visitors, but their duty was entirely environmental.

2.

Visitor dynamics. The Commissioners discussed certain characteristics of tourist behavior but explicitly state this is "a matter not directly comprehended in the instructions of the Commissioners," which was to correct scenic disfigurements. They addressed this subject due to an extensive "public concern."

First they expressed surprise that the, at that point, still healthy climate of the City of Niagara Falls, its high quality inns, and related services and natural attractions should not be "the temporary residence of great numbers of those who every summer migrate from town to country, and one of the most popular places of vacation sojourn in all the world."

They observed that Niagara has no such summer population, but all visits are short. This was not because of harassment, so frequently touted as the cause, nor a failure on the part of the respective municipalities to control the catch-penny tourist trade—these harassments are also characteristic of some of the most famous tourist attractions in the world, indeed, they experience more predation than at Niagara Falls.

Yet tourists at Niagara are "ill-humored" after their visit for some additional reason, which makes the beggars that appeal to them more intolerable than in tourist sites elsewhere.

The source of the quickness of the stay, and the ease with which disappointment is stimulated was considered due to the nature of the landscape itself. This observation is critical to an understanding of the "best use" of Niagara Falls as a tourist attraction and fully understood by the Commissioners, Gardner and Olmsted. It is one that has been paid little if any attention at all by a century of concerns interested in a tourist industry at Niagara Falls. It is a concern that will be more fully addressed by the present writer in the section on recommendations below.

Flat, glaciated terrain of little note extends right up to the crest of the Niagara Gorge. Only several city blocks away from the falls and gorge there is no evidence for the existence of the falls, except, today, for a sound that could be that the railroad yards, and an atmospheric cloud that could be taken for vented steam from a factory stack. This may account for the subsequent century-long attempts by various proposers to build up the expectations of the visitor as they cross miles of unremarkable terrain to prepare them for the spectacle they have come miles to see. Unprepared, the visitor may be unable to easily assimilate the extraordinary phenomenon presented and be "unimpressed," that is to say, unable to shake the boredom associated with the monotony of the adjacent and surrounding prairies and lake plain, or, upon leaving, be unable to retain the excitement they did permit themselves—as was the experience of Judge George Clinton when he saw the falls as a youth of eighteen. Topographically uninteresting terrain leading up to the falls, in the present day, for example, exacerbates this lack of interest by the urbanization on all approaches, which intensifies the sameness of visitor experience and deadens their immediate memory. Similar phenomena probably occur wherever the significant natural feature is a canyon worn below the surrounding regional land surface—even the Grand Canyon which also, at least along the south rim, is invisible from the road paralleling it only a few feet away from the lip of the precipice. Elevations, such as mountains, which can be viewed for long distances before they are reached, can stimulate the visitor upon approach and departure in a way that depressions, such as canyons, cannot.

These sentiments were expressed over a century ago by the famous English geologist Charles Lyell (1845):

"In the region between Lake Erie and the borders of Pennsylvania, as well as in that immediately south of Lake Ontario, there is an entire want of fine scenery, as might have been anticipated where all the strata are horizontal. The monotony of the endless forest is sometimes relieved by a steep escarpment, a river with wooded islands, or a lake; but the only striking features in the landscape are the water falls, and the deep chasms hollowed out by them in the course of ages. As the opposite banks of these ravines are on the same level, including that of the Niagara itself, we come abruptly to their edges before we have any suspicion of their existence, and we must travel out of our way to enjoy a sight of them."

As early as 1819, the educated Scotch botanist John Goldie also noted the strange and, to him, unexpected, flatness of the region approaching the falls. "On approaching them I found the ground in their vicinity to exhibit a very different appearance from what I had expected. Instead of high rocks & precipices above the Falls, and low valleys & glens below them, all is perfectly level to appearance. ... At the distance of 200 yds there is nothing to be seen in the banks of the River that would lead you to expect any such thing as Falls at this place" (Goldie, 1819). Goldie is a perfect example of the nature of the ill-humor, or disappointment frequently remarked upon by proponents and developers of the falls alike. Their expectations of Niagara embedded in a terrain of monumentally deformed geology are shattered. If the wind and air pressure is not exactly right at the moment a visitor gets to the falls, no enormous plume of mist is evident, no thundering roar if the wind blows in the wrong direction, no drenching spray. Every attempt by city, state and province to make inaccessible the singular and concentrated natural charms of the cataracts, for example by permitting helicopter rides whose noise obliterates the sound of the falls, heightens the disappointment factor of the visitor.

The Commissioners had a remedy for this disappointment—to slow the pace of the visitor long enough for the beauties of Niagara's concentrated natural beauty to express themselves. Otherwise, the nearby cities were "well adapted to the bare satisfaction of curiosity in the waterfall."

According to the Commissioners, "the value of Niagara to the world, and that which has obtained for it the homage of so many men whom the world reveres, lies in its power of appeal to the higher emotional and imaginative faculties, and this power is drawn from qualities and conditions too subtle to be known through verbal description." This value can only be appropriated by slowness, by activities and environments promoting a "composed, receptive and contemplative frame of mind." This concept would later be translated into modes of transportation, and later still be abandoned by the administration of the park eventually established at Niagara Falls in the promotion of the automobile and its potential for supporting "the bare satisfaction of curiosity in the waterfall."

3.

The Commissioners resumed their environmental conclusions and determined that the "scenery of Niagara Falls has been greatly injured, that the process of injury is continuous and accelerating, and that, if not arrested, it must in time be utterly destructive of its value." The source of the injury was its then present ownership and the decision by that ownership to "strive to make his particular ground yield the largest possible private profit." The Commissioners determined that since the area on which Niagara's charms are concentrated is so small, that any one attack on it has a "fatal [effect] upon its character."

4.

The Commissioners recommended the employment of eminent domain to condemn the private properties for the public good, such as "when private ferries are supplanted by free public bridges." This was to reserve such land "to give satisfactory access to the Falls of Niagara and preserve their value."

Based on determinations by Gardner and Olmsted, they recommended that the islands in the river and a bit of land along the riverbank on the mainland be taken for the public good.

It is clear what the Commissioners accepted were Gardner and Olmsted's intentions, that is, their "design":

a. the buildings in place were to be removed.

b. the immediate bank of the river "shall be formed so as to have a natural aspect."

c. Its shoreline was to be protected from erosion by modifications of the slope and "by rough, loosely piled local rock."

d. "Trees and bushes are proposed to be planted":

1. composed of the same species as were native to the area and

2. are to be laid out to conform to the ecosystem structure observable in the unaltered woodlands and riverbanks in the Goat Island complex, that is to say "of such kinds and in such dispositions as are natural to the locality."

e. This ecosystem restoration was to be carried back from the river margin "back to the boundary on the crest of the terrace."

f. The effect was to visually isolate the lands in question completely from the buildings of the village of Niagara Falls. It was to "secure their landscape disconnection with the river."

g. Inside of this "narrow woodland," and "along the rear" of it, about "one hundred feet distant from the water's edge" was to be a road and a walk. From the walk would be constructed "inconspicuous shaded seats commanding views of the rapids" except at Prospect Park where there was to be a "more extended platform overlooking the falls and chasm."

The illustration preceding this discourse appears to clearly express the visual objectives of Gardner and Olmsted's recommendations.

5.

The Commissioners made it clear that they accepted Gardner and Olmsted's recommendations as to the limits of the property required to protect the falls, and to the nature of the restoration and preservation of the environment. They concurred with Olmsted, out of "cordiality," as to the use of the reserved land, but "it is not the duty of the Commissioners to advise how it shall be used."

Olmsted urged that the land:

a. not be used for "general pleasuring," in today's terms, such as picnicking, playing tennis, baseball, frisbee, concerts, helicopters, hot-air balloon rides, roller skating

b. not be used for ornamental displays, such as gardens (the artificial laying out of exotic species in unnatural arrangements) and monuments.

c. whatever was to be built was not to break the visual integrity of the natural landscape.

Mr. Olmsted's reasons were based on the understanding that greater numbers of people would eventually have the means to visit the falls. If any one of the preceding three requirements were neglected, such visitation would overwhelm and destroy the natural environment. It was to negate the impact of great visitor numbers that Olmsted made those recommendations.

To protect the considerable value of the native landscape and the landscape restored, the only service the government would be required to provide would be to allow visitors, in comfort and safety, to come, look, and go away. This may seem like an unnaturally proscribed service to tourists, but the visual environment at Niagara is so rich and awesome, and confined to such a small area, that this use was adequate. Besides, that is what visitors were doing anyway, coming, looking, and quickly departing. Olmsted merely wished to protect the integrity of the object of their experience.

This explains the other illustration included in the special report on the preservation of the scenery at Niagara Falls: "View in the primeval woods, on Goat Island," drawn by Thomas Moran and en-graved by a Mr. Karst (plate X). This plate (see page 6) is not simply a pretty decoration to the text, it illustrates the proper relationship between visitor and scenery (environment)—the scenery to be protected and restored. The walker is solitary and enclosed by a rich and complex forest ecosystem. The visitor is walking, not riding, he is strolling with a cane to assist him in his movements and is completely immersed. The path is a dirt path. The forest was actually so dense at this time that a series of such paths could be built into the forest and walkers would be hidden from one another.

The nearby commercial enterprises in the village were adequate to attend to all other visitor needs, hence Olmsted saw no need for "houses of refreshment, shops, booths, and places of amusement and exhibition." Nor did he see any need for "extensive shelters."

Olmsted did, however, entertain the idea of structures built to improve the ability of visitor-access to views. Visitors came from around the world to do what they could do nowhere else in the world: look on the Niagara River at this place. Olmsted suggested to the Commissioners that "at one or two points something might be gained by the erection of belvederes or prospect towers," perhaps reminiscent of the world famous Terrapin Tower the Porters had taken down as part of the bargain for selling Prospect Park to the Prospect Park Company (Porter, 1900). However, Olmsted was not optimistic about the ability to build something inconspicuous nor commodious enough for the thousands of anticipated visitors.

6.

The next paragraph makes reference to Schoellkopf's hydraulic canal. The Commissioners assure the Legislature that it is fully operational. The Porters, who operated the mill on Bath Island, informed the Commissioners that indeed, the hydraulic opportunities on the canal, their "mechanical advantages," were superior to those on the island or "any upon the ground of which it is proposed that the State should take possession." "This would be equally true as to any considerable industrial undertaking"—perhaps a reference to the small scale of the mills already in operation on the riverbank, and the relative insignificance of their displacement. "The provision thus secured can be enlarged, should this ever be required, to any desired extent, and the water power of the falls more economically utilized than if their immediate banks were to be occupied by factories." I interpret this to mean that the canal can be enlarged. Note how Schoellkopf's operations could benefit from the State shutting down milling on the river bank by relocation of milling operations to his hydraulic canal.

The reason why these milling operations occurred between Port Day and the brink of the American Falls was because they were taking advantage, not of the 200-foot head accessible at the Canal Basin, but the rapid 50-foot drop in elevation of the river associated with the ridges (The Cascades) perpendicular to the river bed forming the western boundary of the Chippewa-Grass Island Pool. The hydraulic canal of 1877 and the one built later just upriver of it both diverted water from the pool, above the cascades, exploiting the hydraulic potential of the drop in elevation of the canal bed and the gorge rim, not that of the river. In a sense, the canal and the discharge races at its lower end, were hydraulic analogues of the natural river and falls.

It could not be determined whether the Porters ever actually did relocate their mill on Schoellkopf's canal, or the fate of any of the other mill operations then operating on the riverbank. But then, as will be discussed below, the Porters may have been more interested in establishing a second, rival, canal east of Schoellkopf's.

7.

The Commissioners determined that the cost of the land recommended by Gardner and Olmsted would not be excessive since the boundary lines "are so laid down as to leave out, not only the principal water works, factories and shops for which the Falls have given occasion" but also the hotels and other public accommodations.

Furthermore, the State would, if following Olmsted's suggestions, not need to anticipate "costly constructions or elaborate arrangements for the entertainment of the public." And as for the potential for corruption (by whom and from where is not mentioned), there would be no licenses or leases "which might be corruptly dealt with."

8.

"Niagara Falls is not simply the crowning glory of the great resources of the State of this class [of natural objects], but the highest distinction of the nation and of the continent."

The continued operation of private enterprise on the riverbanks at Niagara Falls was noted to reduce the overall tourist revenue of the State, the restitution of which would outweigh the depletion of the State treasury through the purchase of those lands.

Future generations were expected to deplore the failure of the State in protecting this distinguished resource. "If we blame the men of a former day for not setting apart when it was the property of the State and might easily have been done, the Falls of Niagara as the Yo Semite and the Yellowstone have in our day ..., then how much more culpable shall we be, who knowing their value and perceiving their certain destruction, still refuse to take the necessary measures for their preservation."

It is significant here to understand that Niagara Falls was ranked both physically and politically with the future national parks of Yosemite and Yellowstone—not, for example, with urban parks such as Central Park in New York City, nor the lovely urban parks designed by Olmsted in nearly Buffalo. As a matter of fact, at a time when the federal government was gaining strength and actively seeking custodianship of areas of significant national landscape, such as Yosemite and Yellowstone, it is curious that the federal government was not sought out to intervene—especially when the political head of federal Canada, not the provincial leader, who was Oliver Mowat, appeared to be interested in the protection of Niagara.

Roper (1973) appears to settle this question:

"In recommending the proposal, Mowat advised that the governments of the Dominion of Canada and of the United States should be regarded as the actual principals. While the rich state of New York could well afford to represent the United State's interest, Ontario had too limited revenues to assume a similar burden in connection with what was a national responsibility. That the preservation of the falls was such [i.e., a national responsibility] was surely clear from the fact that the Niagara River was navigable water under dominion control and an international boundary" (from a copy of a letter Oliver Mowat wrote to [Lord Lorne], 9 December, Box 31).

The Canadians understood that it was to be the Dominion Government, rather than the Province of Ontario, which would foot the bill for botanical restoration "since it" rather than the Province "claimed jurisdiction over most of the lands involved" (Way, 1946). Note that some in the State of New York did not think the State could adequately represent the United States in this issue and opposed passage of the 1885 bill in the Legislature establishing the Reservation in law (Welch, 1903).

9.

The Commissioners held a conference with the "members of the ministry of the Province of Ontario in September last," with Oliver Mowat as Prime Minister of Ontario.

They met on September 27 in company with Gardner and Olmsted. "'The general outlines of a scheme which I presented was fully approved by all,' Olmsted wrote [Charles Elliot] Norton. Oliver Mowat ... and one of the commissioners, reported on it to Lord Lorne, successor to Lord Dufferin. Only those arrangements were to be made that were necessary to restore and preserve the natural character of the scenery; it was not intended to make a park or artificial enclosures. The reservation [for State and Province] was to include the islands above the falls and a strip on either side of the river, wide enough for planting to screen out the buildings behind it, from the head of the rapids downstream to the railroad suspension bridge. A modest fee, to defray expenses, would probably be levied on sightseers" (Roper, 1973, from a copy of a letter Oliver Mowat wrote to [Lord Lorne], 9 December, Box 31).

After the report of the Commissioners was written, they received news that "the legislature of Ontario has taken preliminary action for the purpose" and noted this in a footnote.

Ontario, however, wanted the Dominion to carry through with the Canadian part of the reservation "under the limitations of their governmental system." A legal, rather than economic reason is referred to (see discussion under number 8, above).

On the New York side, "it is not necessary to point out the respects which would make it unsuitable for New York to appeal to the Federal government to relieve her from whatever expense the matter may involve. It is sufficient to say that many considerations of State pride as well as of constitutional difficulty, make it clear that if the American part of this work is to be done at all, it must be done by New York alone."

This insistence on the exclusion of the federal government from protection of Niagara Falls, even in the context of federal protection of other national natural treasures such as Yosemite, has been repeated for decades by the Niagara Reservation administration, but the present writer has never found any other reason for it than some reference to "State pride."

The Commissioners reassure the Legislature that the Canadians would cooperate in the legislation to make an international park. In fact, an Act passed "by the Legislature of Ontario in March, 1880, entitled 'An Act respecting Niagara Falls and the adjacent territory'" (Statute of the Province of Ontario, 43 Vict., cap. 13) stated that Canada would, in co-operation with the State of New York, "restore the scenery about the Falls to its natural condition and at the same time afford travelers facilities for observing the points of interest in the neighbourhood" (Way, 1946).

10.

The Commission recommended that the Legislature "take such action ... to acquire the lands" under discussion, and to appoint a Commission to deal with the legal matters in so doing.

**The Report of Mr. Gardner to the Commissioners
of the State Survey**

Gardner began by emphasizing Lord Dufferin's role in initiating legislative action to protect the falls at Niagara, being "first looked upon rather as an expression of philanthropic sentiment than an earnest proposal of a practical measure." That Dufferin may have been "unduly moved" by witnessing visitors persecuted by the peddlers and beggars rife at the falls was considered a motivation with which his critics had no sympathy. Obviously to move to create a reservation required more profound motivation than that, perhaps it being presumed that people could take care of themselves after all. Governor Robinson's motives were environmental: he "appealed to the pride of the people to protect this great and beautiful gift of nature from being degraded into a show ... while the shores, once forest-clad, became mill-sites and places of amusement."

Gardner defined Niagara Falls: the rapids (cascades), the islands in the river, the falls and the plunge pool. He emphasized the quality of the remaining woods and the quality of the experience still to be had, but threatened by the imposition of development: the mill on Bath Island.

Verbally Gardner contrasted the positive and the negative:

Positive: picturesque clusters of evergreens, rich overhanging foliage, deep woods seclusion, surrounded by the influences of nature, graceful woods, banks rich in verdure and overhung with stately trees, pebbly shores, graceful ferns, trailing vines, a mighty torrent writhing and foaming in fury.

Negative: paper mill, started in an "evil hour," unsightly sheds and buildings, disfigurations by wing-dams and ice barriers, the mill an abomination, blank stone walls with sewer-like openings through which tail-races discharge, timber crib work, advertisements, ranks of buildings in all stages of preservation and decay, hotels, mills, carpenter shops, stables, bazaars, ice-houses, laundries, bath houses, rookeries, fences, patent medicine signs, ruin, confusion, solid ugliness.

Later, the State would acquire the "Tugby Bazaar building, the brick and stone shops, the pulp mill and machinery, and the Rapids Hotel building," and sell at least six frame buildings, a planing shop, flouring mill, boat house, bath house, stone foundry, barn, shed, stone house, wing of a hotel building, ice house, store, shop buildings, mill flumes, old fences and lumber. These were "three dwellings, four mills, two hotel buildings, two stores, five stables, two ice houses, one stone house, one pump house and one bath house, beside a number of sheds, and many fences" (2 Ann Rep Comm, 1886). One of the mills was the Witmer mill, and one of the sales, later in 1886, was an "Edison electric light plant" (3 Ann Rep Comm, 1887).

Gardner naively assumed the permanence of the liquid landscape: "The Falls themselves man cannot touch." This is because the hydraulic canal was barely operational and had not yet tested its ability to divert water from the river. Perhaps its owners had not yet computed to the square foot the amount of water it could rent to those using water in their mill sites situated on the canal. Gardner confined his discussion to the destruction of the "beautiful frame of foliage" at the Falls.

Wooded character — Loss of foliage creates "deep feelings of regret and even of resentment" on intelligent visitors. "The chasm below the Cataract depends for its impressiveness largely upon the wooded character of the debris slopes and the maintaining of a fringe of verdure along the very brink of the precipice." These elements are "essential to the perfection of the landscape." Note that in the heliotype print of the Canadian side as seen across the brink of the Horseshoe Falls how destitute of vegetation is the bank of earth behind the buildings in the foreground. Today, this embankment is heavily wooded with trees of considerable age. Note, except for the canopies of scattered trees down on the embankment, the complete lack of trees on the top of the embankment.

The only acreage left with its original forest was Goat Island—if the State did not purchase it, even this would be lost. The Porters had to and were in the very process of selling it "owing to a partition suit now in progress." Gardner gave a list of developments proposed for the island, no doubt prepared, according to subsequent literature published by a member of that family (Porter, 1900), by the family itself.

Note at this juncture how helpful the Porter family was being in assisting the Commissioners, Gardner and Olmsted in preparing arguments in favor of the State buying their land, first in the case of their milling operations on Bath Island, as discussed above (section 6), and now in the case of Goat Island itself. It appears as though the Porters were leaving their water-front properties at the brink of the American Falls—but perhaps not those upriver of the entrance to Schoellkopf's hydraulic canal.

Restoration — Gardner again specified the environmental nature of the State's role: "to restore to all the river shores something of their original character."

Gardner cited several examples of government setting aside land for public protection: the Yosemite valley, 1865, the "great tract covering the region of the Yellowstone Geysers—a National Park, the land occupied by the California Big Trees, and, in New York State, the Islands of Lake George."

Gardner chastised the State for selling any of the five-mile strip of land along the Mile Strip embracing the falls of the Niagara River. He urged that "the spot [be] restored by planting to its former beauty," that the mainland strip be planted with trees so that "the whole village may be shut out from view—'planted out'" and the banks restored to their present appearance on Goat Island.

The utility of including both the drawing described above, showing the restored view, and the heliotype prints showing the developed condition of the riverbank was explained: "To realize the total change that the carrying out of this plan would make in the aspects of Niagara, those who are not familiar with the scene may compare the accompanying photographs of the village shore with the picture of the same ground *as it will appear when restored according to our plan* [my emphasis].

These illustrations were not decorations but served a definite purpose.

Gardner recommended that the State also purchase the debris slopes on the mainland section "for the purpose of preserving and restoring the woods that border this part of the river."

"We also recommend that the right be secured to plant and maintain a narrow belt of trees with a walk at least a mile in length along the edge of the cliff below the suspension bridge. This planted belt need not be over twenty five or thirty feet broad. Its trees will clothe the barren nakedness of the cliff edge and partially screen out mills and unsightly structures from the river views, and at the same time afford shade to visitors enjoying the profound impressions of this part of the chasm." The State need not buy the land but only secure a right to plant and preserve. The property belongs to the Hydraulic Power and Canal Company and is to be used for Mills. The walls of these mills will be set back from the cliff, their wheel pits only being sunk at the edge of the precipice. There will be few of these pits, and they can be easily bridged for the proposed walk. The President of the Company owning this property [Jacob Schoellkopf, according to Adams, 1927] has assured us that he will willingly cede the desired right to the State."

Gardner urged that a board of commissioners be set up to assess the value of lands slated for condemnation under the right of eminent domain.

Visitors — Prior to the era of the railways, only the rich enjoyed Niagara. Now everyone could. A total of 100,000 people visited the falls in 1897. Ownership by the government would make the area open to all citizens.

Again, Gardner urged the illustration showing the objective of his and Olmsted's design be examined: "Although truthful in the general impression conveyed, such a view cannot, of course, be accurate in detail."

The Goat Island forest is a living monument of history, and so facsimiles of Hennepin's description and illustration are included in the report, "this first recorded visit of a white man to the Falls." The ancient trees on Goat Island have witnessed all the history of the past two hundred years, they are "the only living witnesses" of the passage of history at the falls and must be preserved. One cannot help but wonder if this interesting historical perspective was provided by the Porters, who obviously consulted with both Commissioners, as noted above, and Olmsted and Gardner. Albert Porter had written a short pamphlet published sometime after 1875 detailing the history of the village and the Porter fortunes there, and some of Gardner's historic sentiments reappear in Peter A. Porter's history of Goat Island published for the Legislature in 1900.

Gardner ends with the declaration of the value of the geologic environment of the Falls for study: "the conclusions to be attained by accurate geological study of the region open almost limitless views into far-reaching vistas of the continent's physical history."

The value of the scenery, associated history and opportunity for scientific study may be preserved by the State against the values of "money-getters," the "axe of the mill-man," the "purveyor of public amusements," that is, the present owners of the riverbank properties.

The Report of Mr. Olmsted to the Commissioners of the State Survey

It was Mr. Olmsted's opinion that "most of the people of Niagara [are those] to whom it appears that the waterfall have so supreme an interest to the public that what happens to the adjoining scenery is of trifling consequence." His opinion derives from personal experience with local opinion. "Were all the trees cut away, quarries opened in the ledges, the banks packed with hotels and factories, and every chance-open space occupied by a circus tent, the falls would still, these think, draw the world to them." This opinion, indicated Olmsted, derived from profit alone as the sole value.

Olmsted cautioned that because visitors use the arrangements made for them, they must be considered a captive market. Their use of the facilities should not be taken as their acceptance of their approval.

Over the course of forty-five years of occasional visiting Niagara Falls, Olmsted recalled a gradual quickening of pace throughout this time. Visitors originally alighted from their carriages and made expeditions into the natural areas over the course of several days. It was because they were hurried along by tourist-related "services" that the duration of their stays decreased.

Olmsted gave an extended quote from "Alpine Flowers" by William Robinson (1875) who provided a lavish description of the natural environment about the Falls. Olmsted identified two of the world's most distinguished students of botanical science, Sir Joseph Dalton Hooker of Kew Gardens, and Dr. Asa Gray of Harvard, as expressing their appreciation of the exotic diversity of the vegetation in the Goat Island complex, and Olmsted, referring to his extensive travels in the American southwest throughout the rich forests of the Appalachian mountains, has found no example of forest beauty to match that on Goat Island.

He owed this extraordinary condition of the flora to its extraordinary situation beside the cataracts of the Niagara River and discussed various atmospheric reasons for the luxurious beauty of the Goat Island forest.

Olmsted ended his section by a quote from the Duke of Argyle who became nearly prostrate with delight by standing in one single vista up the rapids from the falls. Otherwise, he rested his case on the statements made by the Commissioners and by Mr. Gardner, made, where appropriate, at his recommendation.

Some interpretations — These are my graphic evaluations of suggestions in the Report.

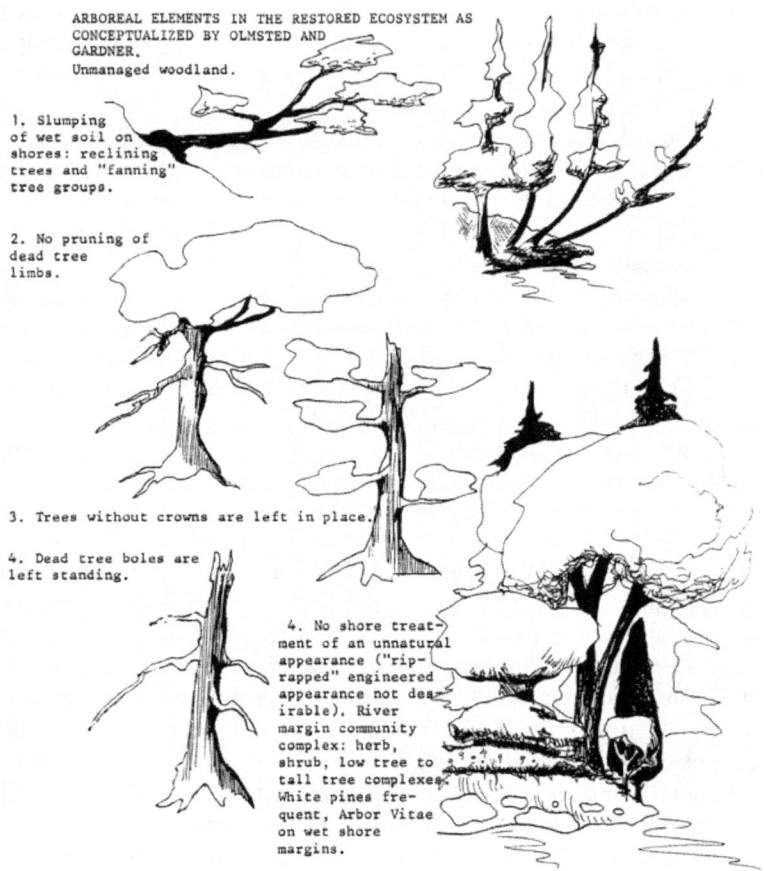

ARBOREAL ELEMENTS IN THE RESTORED ECOSYSTEM AS CONCEPTUALIZED BY OLMSTED AND GARDNER.
Unmanaged woodland.

1. Slumping of wet soil on shores: reclining trees and "fanning" tree groups.

2. No pruning of dead tree limbs.

3. Trees without crowns are left in place.

4. Dead tree boles are left standing.

4. No shore treatment of an unnatural appearance ("rip-rapped" engineered appearance not desirable). River margin community complex: herb, shrub, low tree to tall tree complexes. White pines frequent, Arbor Vitae on wet shore margins.

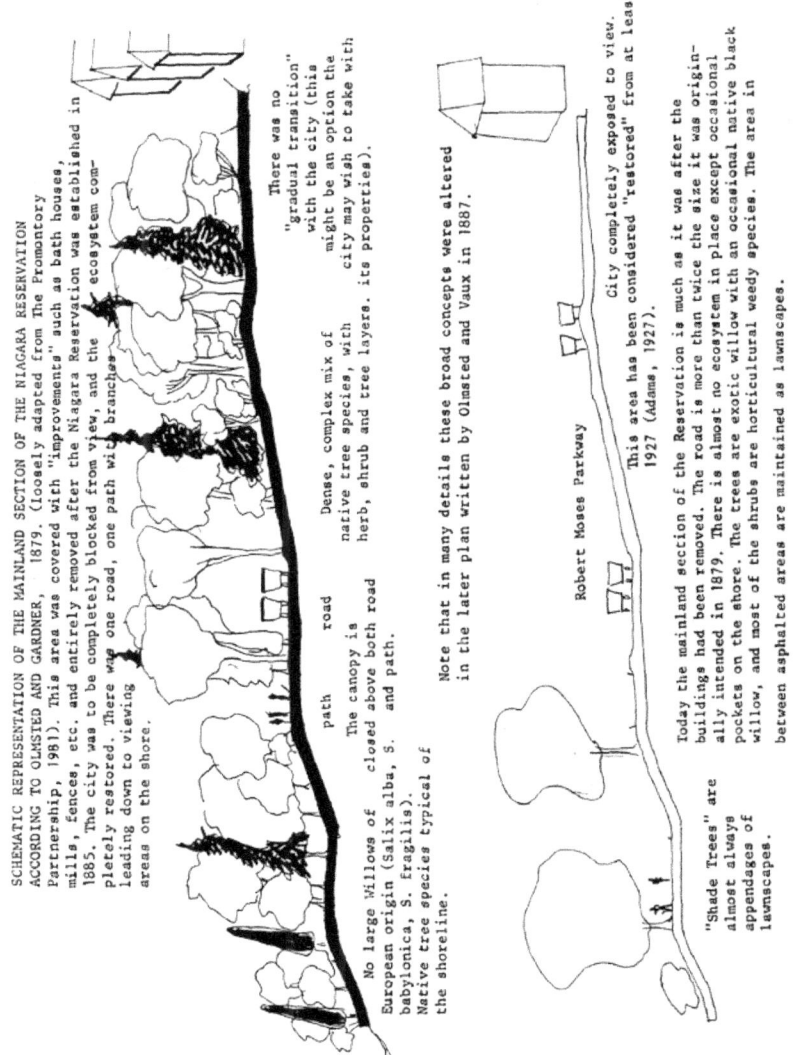

SCHEMATIC REPRESENTATION OF THE MAINLAND SECTION OF THE NIAGARA RESERVATION ACCORDING TO OLMSTED AND GARDNER, 1879. (loosely adapted from The Promontory Partnership, 1981). This area was covered with "improvements" such as bath houses, mills, fences, etc. and entirely removed after the Niagara Reservation was established in 1885. The city was to be completely blocked from view, and the ecosystem completely restored. There was one road, one path with branches leading down to viewing areas on the shore.

There was no "gradual transition" with the city (this might be an option the city may wish to take with its properties).

path   road

The canopy is closed above both road and path.

Dense, complex mix of native tree species, with herb, shrub and tree layers.

Note that in many details these broad concepts were altered in the later plan written by Olmsted and Vaux in 1887.

No large Willows of European origin (Salix alba, S. babylonica, S. fragilis). Native tree species typical of the shoreline.

"Shade Trees" are almost always appendages of lawnscapes.

Robert Moses Parkway

City completely exposed to view. This area has been considered "restored" from at least 1927 (Adams, 1927).

Today the mainland section of the Reservation is much as it was after the buildings had been removed. The road is more than twice the size it was originally intended in 1879. There is almost no ecosystem in place except occasional pockets on the shore. The trees are exotic willow with an occasional native black willow, and most of the shrubs are horticultural weedy species. The area in between asphalted areas are maintained as lawnscapes.

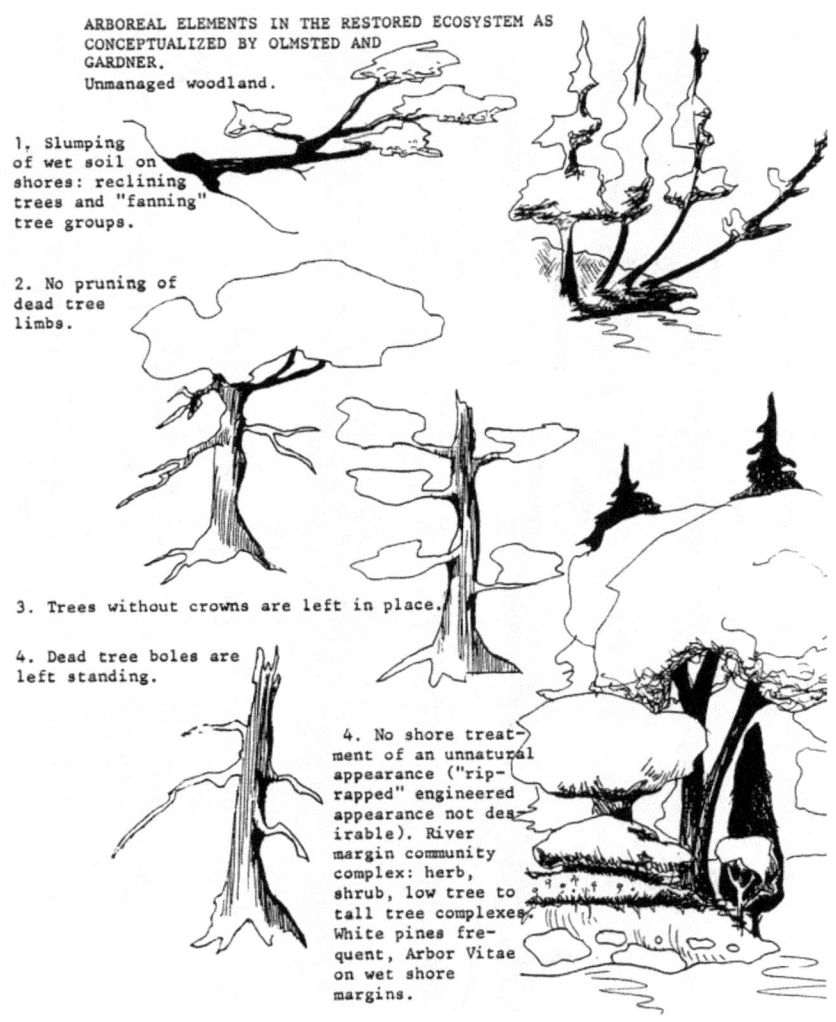

ARBOREAL ELEMENTS IN THE RESTORED ECOSYSTEM AS
CONCEPTUALIZED BY OLMSTED AND
GARDNER.
Unmanaged woodland.

1. Slumping
of wet soil on
shores: reclining
trees and "fanning"
tree groups.

2. No pruning of
dead tree
limbs.

3. Trees without crowns are left in place.

4. Dead tree boles are
left standing.

4. No shore treat-
ment of an unnatural
appearance ("rip-
rapped" engineered
appearance not des-
irable). River
margin community
complex: herb,
shrub, low tree to
tall tree complexes.
White pines fre-
quent, Arbor Vitae
on wet shore
margins.

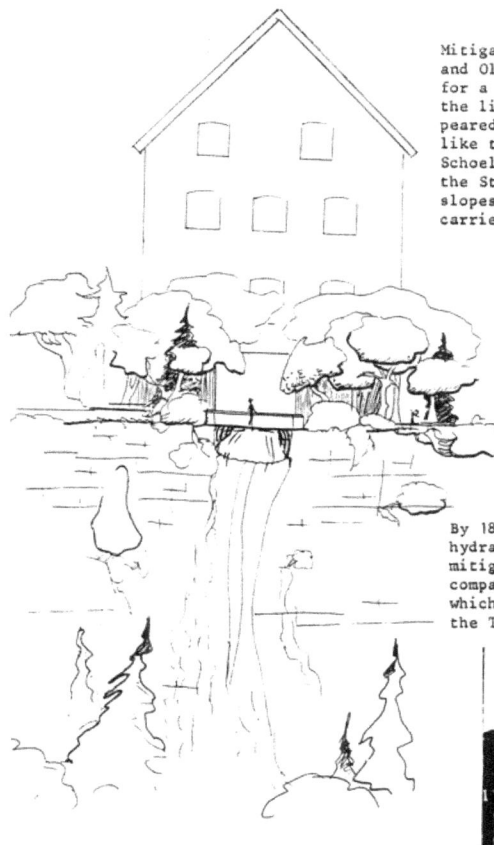

Mitigations suggested by Gardner and Olmsted to Jacob Schoellkopf for a stip of land running along the lip of the gorge may have appeared in imagination something like this. Although in 1879, Schoellkopf was willing to allow the State to restore the lower slopes and the rim, this was not carried out in subsequent years.

By 1893, the lower end of Schoellkopf's hydraulic canal showed no attempt at mitigating the visual impact of the companies using his canal, a situation which called for remedy later by the Taft Commission (see text).

(photo: Adams, 1927).

The "Memorial Addressed to the Governor of New York and the Governor-General of Canada"

This is the result of a petition. Lord Dufferin is given exclusive attribution for the political suggestion that there be an international park at the falls of Niagara.

The undersigned universally deplored the de-foliation of Niagara's river banks and the development arising in their place.

There followed a list of the world's luminaries in society and culture, including the Vice-President of the United States, the Chief Justice of the Supreme Court and perhaps all of the Associate Justices, the Chief Justice Court of Appeals, Canada, the Judge of the Queen's Bench, United States Senators, an Admiral, and so forth.

Then follow English men of letters such as Thomas Carlyle, John Ruskin, and famous Americans such as Ralph Waldo Emerson, Henry W. Longfellow, and so forth. It is an extraordinary list of contemporaries of the greatest prestige. Western New York's own Josiah Letchworth, George W. Clinton and William Dorsheimer appear, and so was the manufacturer Pascal P. Pratt, who sat on the Board of Parks Commissioners and was central to the plan to bring Olmsted to Buffalo to distinguish the city with a series of urban parks under his design (Brown & Watson, 1982).

The composite report concluded with a facsimile of the pages of Hennepin's book with his description of the falls, and a reprinting of Governor Robinson's message to the New York State Legislature of January 9, 1879.

POWER DEVELOPMENT IN THE 1880'S

While the tourist industry at the Falls was making life unpleasant for visitors throughout the 1870's and into the '80's, Schoellkopf's successful development of the hydraulic canal began to change the community. Unfortunately, such development also lead to defacement of the previously intact wall of the American side of the gorge. The mill of Charles B. Gaskill, erected in 1875, broke the line of the gorge rim and rose up over its discharge race which erupted from the 200 foot gorge wall twenty five feet below the rim. Its cascade was the second, matching the then unutilized water pouring over the rim from the canal itself (Adams, 1927).

Charles B. Gaskill, 1841–1919 — President, Niagara Falls Power Company, 1886–1894; Colonel United States Army, served with distinction in Civil and Spanish-American Wars (Adams, 1927).

Schoellkopf and associates advertised for sale one hundred additional mill and factory sites and three hundred cottage lots (Adams,

1927). Schoellkopf's Niagara Falls Hydraulic Power and Manufacturing Company was incorporated in 1878. The second mill, also a flouring mill, was erected by the firm of Schoellkopf and Mathews "between the canal and the top of the cliff." A third mill, also owned by Schoellkopf under the name of the Central Milling Company was erected "northerly of the other mill," then the "paper industry was the next manufacturing interest to utilize the canal. ... This naturally attracted and promoted the location of wood pulp factories and the manufacture of paper products" (Adams, 1927). By 1882 there were three pulp mills, two flour mills (the third must have been erected after 1882), a silver plating manufactory (Oneida Community, Ltd.) and the village water works for the town of Suspension Bridge (Adams, 1927). In the latter instance, the company played a community role, sharing its economic plant with a community necessity—much like a utility. As a result of these industries and the labor they employed, "the population of the village of Niagara Falls increased" in the early 1880's.

Mills and power plants had to be built close to the cataracts both above and below their brinks to take advantage of the drop in elevation and because of the limitations of the hydraulic technology of the time. Lumber yards and mills were mixed with hotels and public houses in general along the banks of the upper river. The shoreline "between Prospect Point and First Street was occupied by mills, bazaars, hotels, stables, ice houses, bath houses, pump houses, laundries, sheds and other structures" (Scott and Scott, 1983).

Tourism versus power — Tourists struggled to see a great phenomenon of nature while indifferent industries blasted tunnels in the gorge walls, built their structures on the river margins and indiscriminately dumped the effluents of their industrial processes and the debris of their operations. "Since there was no machinery at that time [mid-nineteenth century] capable of withstanding a "head" [actual or potential vertical distance of falling water] of more than twenty-five or thirty feet, the water wheels were installed in a pit of about that depth and the water after use was spilled over the cliff into the river, with a waste of more than 100 feet of "head," while the power generated was applied directly by shafts and pulleys to the mill machinery" (Way, 1946). The Niagara Falls Hydraulic Power and Manufacturing Company, which operated the hydraulic facilities in the Canal Basin, or Mill District of Niagara Falls, as the lower end of the hydraulic canal was called where all Schoellkopf's mills were located, was the enterprise dumping waste water out of their client's pits in the gorge wall above the American Falls.

Now, not only was the view required to be bought by visitors at considerable expense and harassment, but it was becoming marred by industrial mutilations.

Hydroelectric power generation — Then, in 1881, "the first hydro-electric generating station was established on the hydraulic basin to supply electricity for commercial purposes" at the Quigley Mill, later the Cliff Paper Company mill. An arc light machine was operated by this mill, owned by the Brush Electric Light and Power Company, a Schoellkopf concern, and a number of applications for the use of the electricity for illumination were submitted. Light was used to illuminate fountains in Prospect Park and to illuminate the rapids from Bath Island. So promising was this demand for what electricity there was that "In 1886, The Niagara Falls Hydraulic Power and Manufacturing Company secured a deed for the slope or strip of land between the high bank and the lower river, that was not included in the original grants acquired from the Porter family" (Adams, 1927).

The great electric utility revolution had been born sometime in the 1870's and had just begun to expand with the movement to "free Niagara." In all the literature generated for the liberation of the cataracts and adjacent river margins from private enterprise, in all the recounting of the actual and potential threats to the beauty of what was to become the Niagara Reservation, there is never any indication whatever that the liquid landscape itself was in danger of disappearing.

Jacob Schoellkopf, his son Arthur and other associates made the hydraulic canal operational within five years of its purchase. In a curiously undated prospectus issued by the Niagara Falls Canal Company, but interpreted by Adams (1927) as issued soon after purchase of the canal and before Schoellkopf erected his own mills, the grandeur of the canal potential was described by the offer of sale to the business community of one hundred mill and factory sites and three hundred cottage lots, presumably for the labor associated with those businesses. The authors of the prospectus were able to calculate to the dollar the value of the water by volume offered to those using the canal. The water "at one thousand dollars per square foot of open weir surface at the head of the canal and the opening in the gates below to correspond in size to the square of water purchased." Furthermore "it was estimated that a square foot of water at the entrance of the canal, which would be one six hundred and sixtieth part of the whole inflow of water, with a velocity of 2 1/2 feet per second, would give 41.67 horse-power, 'a much more liberal estimate for loss of power upon water-wheels than is generally allowed' so the prospectus claimed" (Adams, 1927). The canal itself "was improved and enlarged from time to time" (Adams, 1927), naturally to capture more water to be sold according to the parameters just noted.

The significance of these figures issued in the prospectus of the Canal Company is that a monetary value could be calculated per unit volume of water, its velocity and the horse-power generated, even down to the calculation of its value with respect to the efficiency of the industrial technology employed by the company to generate power. An entirely practical reduction of the great world-class wonder of nature was now possible by a simple application of these figures to the river itself. The river, after all, was just a natural canal or mill race. Once the dimensions of this great race could be established, the volume of water in it, its velocity, and minor details such as seasonal variations and problems with ice jamming equipment and blocking feeder canals in winter, the entire amount of water flowing over the cataracts could be systematically reduced to dollars and cents, and a price placed on all the water falling "unutilized" over the brink to the thriftless delight of millions of visitors.

The prospectus also indicated that its authors were fully aware of the transformation of the adjacent community that the success of their venture would entail—a community nearly fully, but not quite, dependent upon the canal operators. The tourist industry would always operate independently of the manufacturing industry flourishing beside the canal.

Proponents of the financial value of the falls and proponents of its cultural and social values each had powerful political and social advocates, private and public. Those focusing primarily on its financial value appear to have been represented by private interests, while the social and cultural proponents went to the public in their operations. It was in the interests of the financiers to downplay the cultural and social value of the falls, or transfer the basis for this value onto man-made rather than natural objects, artifices whose production would not have been possible were it not for the awesome power of the water. Those who saw the social and cultural value inherent in the falls insisted on its untrammeled, unchanged natural character to embody those values. Naturally, the more alterations in the natural character of the Niagara environment, the more diminished the social and cultural values, and the strengthening of the financial, or economic ones.

Antagonistic to the financial reduction of the value of the falls of Niagara would be the tourist industry itself. Such an industry made its bread and butter out of the curiosity of travelers, and required a waterfall with some beauty attached to it as the basis for its market. It is easy to see a natural conflict between control of the economy of the developing community of the City of Niagara Falls by a manufacturing industry and a strong tourist industry with a high tourist clientele. These tourists formed the constituency of a political movement that would cost the commercial sector its economic investments along the riverbanks, tourist and manufactory alike.

Interactions of power development and the Reservation — It is important to consider what was occurring on the hydraulic canal during the decade of the political movement to establish the Niagara Reservation. Oddly enough, the elegant history of the Niagara Power Company written by Edward Dean Adams (1927), who is extremely detailed and scrupulous about the developing power industries of the period makes a hiatus in the development of the canal, leaving a gap from around 1882 to 1893, jumping in his history essentially from the first electric power generation at the Quigley Mill to construction of the Schoellkopf power stations in the area in or adjacent to the Milling District on the Niagara River Gorge in 1895 and '96. The impact of the momentous events of the intervening years on the developing power industry on the hydraulic canal is not discussed. Suddenly in 1886, the second great power company, the Niagara Power Company, according to Adams, 1927, literally leaps into conception and its eventual success is detailed. The interaction between both of these power companies and the establishment of the Niagara Reservation is not discussed, nor are mentioned the Canadian power companies established within Canadian park boundaries within the first ten years of the establishment of these "reservations." As it would happen, almost in the same instant the parks were established, the power companies were as well, in Canada within park boundaries, in the United States forming the upper and lower boundaries of the Reservation to be. They would begin to immediately siphon off water from the river, water whose value could be calculated in dollars per volume. The tourist industry would suffer a setback: the loss of major and minor establishments built to serve tourists, also lost were all the independently operating mills on the river bank up to the boundary of the inlet of the hydraulic canal (Port Day). The Adams power generating plant and inlet canal would be established on the river margin immediately east of the canal.

The Movement to Free Niagara

Returning to 1880, with the submission of the report of the New York State Survey on the preservation of Niagara Falls, a bill was introduced in the Legislature which failed in the Senate. In the City of Niagara Falls "increasing manufacturing and increasing demand for water power sent the price of land around the falls edging rapidly upward; and sharp opposition to the park project was mounted by the Niagara Falls Gazette and by Hall and Murray, proprietors of a big wood-pulp mill, who owned seven hundred feet of the coveted shorefront" (Roper, 1973).

Another bill was introduced in 1881, "but no effort was made to secure its passage" (Report of the Executive Committee of the Niagara Association, 1885).

In 1882, Schoellkopf's canal, as discussed above, sported at least five mills, a silver plating manufactory and a village water works. The canal was obviously a success and the businesses on the mill sites presumably were, too. No bill was introduced in the Legislature in that year, but the Niagara Association was created in New York City with the express purpose "to promote legislative and other measures for the restoration and improvement of the natural scenery at Niagara Falls, in accordance with the proposed plan of the Commissioners of the State Survey" (Report of the Exec. Comm. Nia. Assn., 1885).

The first meeting to organize this society was called by a friend of Frederick Law Olmsted's, who later became the society's president: Howard Potter. In a footnote, Roper (1973) described Potter: "(1826-97) son of Bishop Alonzo Potter of Pennsylvania, was a member of the New York bar, but by this time had quit practice to become secretary and treasurer of the Novelty Iron Company in which his father-in-law, James Brown, of Brown Brothers, had the controlling interest. Later he joined Brown Brothers and became a partner in 1861." In 1881, Mr. Potter, attending the Bankers' Convention meeting at the Falls, spurred its participants to approve "a resolution prepared by Norton and Olmsted endorsing the park plan" (Roper, 1973). The organizing members of the society meeting at Mr. Potter's house at his invitation (Roper, 1973), in 1882, included Olmsted, Dorsheimer, Harrison, Norton "and others." There was a "large attendance and much enthusiasm at this meeting, as it was understood that Governor [Grover] Cleveland, who had just been elected Governor of the State [in November of 1822,] was warmly in favor of saving the scenery of the Falls, and would approve of any measure tending to this result" (Report of the Exec. Comm. Nia. Assn., 1885).

The express purpose of the Association "shall be to promote legislative and other measures for the restoration and improvement of the natural scenery at Niagara Falls, in accordance with the proposed plan of the Commissioners of the State Survey, as presented in their special report on the subject, under a concurrent resolution of the Legislature of the State of New York, May 19, 1879" (Report of the Exec. Comm. Nia. Assn., 1885).

According to the document just cited, an attempt would be made to appeal to the State for the outright purchase of Goat Island, and the property about the Falls. This idea of an "outright purchase" is rather confusing because it implies there had been another way to acquire the required properties which had failed due to "the difficulty of meshing New York's moves with Canada's" (Roper, 1973). The outright purchase also, for some reason, required the abandonment of the idea of an international park, hence the deletion of Lord Dufferin's involvement in the Association's reprint of Robinson's message to the Legislature in 1879, as discussed above.

The purchase — On January 30, 1883, a bill was introduced and passed in March of that year, signed by Governor Cleveland, which authorized the "selection, location and appropriation" of land in the village of Niagara Falls, New York "for the purpose of preserving the scenery of the Falls of Niagara, and of restoring the said scenery to its natural condition." Governor Cleveland would appoint five Commissioners, The Commissioners of the State Reservation at Niagara, for five-year terms, unpaid; they were to condemn the respective properties. A Commission of Appraisement would set the actual reimbursements. Passage of this 1883 bill was just as momentous as the final bill of 1885. The President of the United States actually wrote in support of the 1883 bill (Report of the Exec. Comm. Nia. Assn., 1885).

Governor Cleveland then appointed William A. Dorsheimer, Andrew H. Green, J. Hampden Robb, Sherman S. Rogers, of Buffalo, and Martin B. Anderson, president of Rochester University, as Commissioners of the Niagara Reservation. Dr. Anderson became its president, although he resigned as president after some months, Dorsheimer assuming that office, and Mr. Robb became Treasurer and Secretary. On June 9, 1883, the Commissioners, Frederick Law Olmsted, Calvert Vaux and James Gardner accompanied the Commissioners to Niagara Falls where "it was decided to carry out substantially the plan originally suggested in the report of the State Survey," that is, Olmsted's plan, and to conduct a survey of the property by the State Engineer, who appointed Thomas

Evershed, Division Engineer of the State canals to accomplish this (Welch, 1903). The lands were condemned and a board of appraisers selected by the Commissioners, granted a total of awards amounting to $1,433,429.50 (Report of the Exec. Comm. Nia. Assn., 1885). The Hon. George Clinton of Buffalo introduced additional legislation in 1884 relating to the appraisement (Welch, 1903).

The Report of the Executive Committee of the Niagara Association ends with the completion of the report of the Commissioners. The final round of legislation in 1885 is detailed by Welch (1903).

When the Legislature opened in 1885, a bill was prepared "for an appropriation to provide for the payment of the awards made for the lands selected" (Welch, 1903). This would be the final action required to establish the Niagara Reservation. The State Comptroller and other officials overseeing the Treasury "did not wish to quit office with a depleted treasury" and were not in favor of the bill's passage (Welch, 1903).

Opposition to the bill came from the granges "and other organizations of farming people," and from several counties, opposition being "especially strong among the rural members" (Welch, 1903).

Cleveland was said by Welch to have made a statement to him and Senator Robb, when approached for the preliminary bill of 1883, "adding significantly that if the Reservation were to be established, the sooner it were done the better, as it could be done much more reasonably at that time than in 10 or 15 years in the future" (Welch, 1903).

To get out of the fiscal impasse, it was suggested to proponents of the final bill to have the State "issue bonds for $1,000,000, payable in 10 annual installments, and to pay the remainder of the awards out of the funds in the Treasury," although this arrangement appears to have been dropped in favor of paying directly out of the State Treasury. Before, however, "A. Augustus Porter wrote favoring the ... bill, providing for the issue of bonds, and urging that the land-owners be given the option to take bonds in payment, saying: 'This course would, I think, be most decidedly agreeable to those who now have land investments which the State proposes to throw into money for reinvestment'." The constitutionality of the suggestion of issuing bonds was challenged by the "counsel for the commissioners," Ansley Wilcox. Apparently the raising of bond money was allowed by the State Constitution only in matters of grave emergency. According to Wilcox, the State had a $2,500,000 surplus, anyway, which could easily cover the appraised cost of the condemned land.

The Niagara Falls Association, still intact, sent its corresponding secretary, Mr. J. B. Harrison, throughout the State "to bespeak the cooperation of the editors of newspapers and magazines, writers, college professors, clergymen, and professional and business men generally" to support enactment of the bill to preserve the environment at Niagara. A letter writing campaign was initiated on a local level, "among prominent citizens of the village of Niagara Falls (Welch, 1903) in which replies were solicited. Additional letters were generated by the Niagara Falls Association authored by Thomas Welch and generally distributed together with a stamped envelope addressed to a legislator.

Those against the purchase — Opposition to the bill came from those among "legislators and ex-legislators" who thought that "the Nation, and not the State, should take measures to preserve the scenery of Niagara," "there was a large and influential body of our citizens who believed the project should be carried out by the National Government." "Others feared it would lead to a public scandal similar to that incurred in the construction of the new [State] Capitol. Still others regarded it as the entering wedge to the gigantic Adirondack Park scheme, which they condemned" (Welch, 1903). Others felt that the bill was a trick propounded by the citizens of Niagara Falls who were characterized as "sharks and swindlers, who had robbed the people individually and were now seeking to rob them collectively" (Welch, 1903).

On February 18, 1885, Senator Robb, one of the Niagara Commissioners, and also a State Senator, introduced the bill. Governor David B. Hill was expected to oppose the bill.

The "large land-owners" objected to the value appraised on their lands, especially as riparian owners, "claiming to own ... the water power of the river" (Welch, 1903). The final owners appealing the decision of the appraisers "were reduced down to Mrs. Burrell, Mr. Hill and the Prospect Park Company" who had purchased the Prospect Park property from the Porters in 1873, and for whom the Porters had torn down the Terrapin Tower on Goat Island because it competed with the views from Prospect Point (Porter, 1900).

Still, "a flood of petitions poured in upon the Senate and Assembly from all sections of the State" (Welch, 1903).

In April 18, 1885, the Governor actually made a visit to the Falls to see for himself conditions there. Bath Island was "almost entirely covered by the paper mill and other buildings used in connection with it. Prospect Park was surrounded by a high picket fence" (Welch, 1903). The Bath Island mill, "with its chimneys, shops' tables, sheds, straw stacks, fences, flumes and piers, was, of all the structures proposed to be taken, the greatest disfigurement of the scenery, because of its conspicuous location in the rapids" (Welch, 1903). The Governor indicated "that he could not understand why the commissioners had included [Bath Island] with its costly paper manufacturing plant, in the territory to be taken for a Reservation," a statement causing alarm among the park advocates.

The Governor mentioned to Herbert Thompson that "the National Government ought to carry out the project" (Welch, 1913).

Oddly enough, even though it was known that, according to a technicality in the 1883 bill, all work done to purchase the land condemned would be nullified if subsequent legislation were not passed "on or before April 30, 1885," "for some reason the bill [after its passage in both houses of the Legislature] did not come into the hands of the Governor until four or five days after its passage" (Welch, 1903). Welch suggested that President Grover Cleveland may have contacted Hill to encourage his signature.

The bill is signed — After frenetic activity to get Governor Hill to sign the bill, he did so only on the very last day.

On Wednesday, July 15, 1885 "the State Park at Niagara Falls was formally delivered to the people of the State of New York" (Welch, 1903).

About 75,000 people showed up at Niagara to celebrate the triumph of the park movement in the State of New York, the "largest gathering ever there up to that time" (Welch, 1903). Two thousand troops were on parade, the United States regulars, the National Guard, and marines from the United States man-of-war, the Michigan. Music was provided by marching bands from Niagara Falls, Buffalo, Utica and Cleveland, in the United Stated, and by the Mexican National band. Cleveland also contributed the Cleveland Grays, a state militia of Ohio. A chorus of 400 was composed from the Orpheus and Schubert societies of Niagara Falls, and the Orpheus, Saengerbund and Liedertafel societies of Buffalo. Addresses were given, among them, from Dorsheimer, president of the Niagara Commission, and Governor David B. Hill. A committee was sent by the Buffalo Historical Society representing many Buffalo men of distinction. Lieutenant-Governor John Beverly Robinson of the Province of Ontario attended, and "many other Provincial officials and officers of the Niagara Park Commission for Ontario" (Welch, 1903). "Letters were read from President Cleveland, the Governor General of Canada, and Samuel J. Tilden: and a cabled message of congratulation from the Commons Reservation Society of London, England. The day ended in a blaze of fireworks glory on both sides of the Niagara" (Welch, 1903).

Among those attending was the Hon. Oliver Mowat, Premier of Ontario, according to Welch, 1903, but at the time, Attorney-General of the Province (1 Ann Rep. Comm. 1885). Mowat was later to be known as the "Defender of Provincial Rights" for his role in "[wresting] ownership of the province's hydroelectric power from private interests after a long struggle in Ottawa" (Kiwanis Club, 1968).

THE OLMSTED-VAUX PLAN OF 1887
(3 Ann Rep Comm, 1887)

The Commissioners now needed a plan to protect and restore the primitive scenery of Niagara Falls. In every instance of legislation up to and including that of 1885, the objectives and conceptual context of the policy to be applied to Niagara was to follow the 1879 survey, articulating environmental concepts formulated by Olmsted and carefully agreed to in all its parts by the Commissioners of the State Survey and James Gardner, its Director.

The publication of The General Plan for the Improvement of the Niagara Reservation "marked the end of [Frederick Law Olmsted's] important involvement in Niagara matters, nearly twenty years after he first took them up" (Beveridge, 1985). Calvert Vaux contributed perhaps certain technical details for specific constructions, but Olmsted commanded the plan. It was to build on the plan of 1879, accepted and articulated by the Commissioners of the State Survey and the Director of that survey. "The main responsibility for writing the report ...fell to Olmsted, and he agonized over it as he sought to clarify the particular problems of the place and to present the reasons for his design. He told a friend that he was more interested in the problem of Niagara than any other living man, but lamented 'I can no more write what is in my mind about it than a crow can sing'" (Beveridge, 1985).

His agony may be attributed to his bitter experience with Central Park, in New York City, the center of his professional life. After Olmsted's dismissal from the direction of Central Park, five incompetent superintendents had been appointed. The fifth "set diligently to 'open vistas'" (Roper, 1973), with the result that Fifth Avenue "which had been carefully screened out with plantations, was seen from within the park in all its architectural elegance ... and interior views, supposed to steal beguilingly upon the eye, were fully exposed to it at a glance," promoters were urging staging of the world's fair in Central Park and "there was a strong effort brewing to place a tract for fast driving in it" (Roper, 1973). Maddened by "incessant pressures, first from Tammany, then from the 'reformers,' to which he had been subject as superintendent to give out

Andrew Haswell Green — "Without his tireless management, skill in navigating municipal and state politics, and undaunted spirit of a do-gooder, New York City would certainly not have Central Park, The New York Public Library, The Metropolitan Museum of Art, The American Museum of Natural History, New York City Hall...the list goes on and on. Often called the Father of Greater New York, Green was until his dying day a relentless advocate for the common good. He earned this title for being the proponent for consolidation of the five boroughs to make Greater New York City in 1898." (Rubbinaccio, 2013).

jobs for reasons of political expediency and the failures and frustrations that ensued when nonprofessional considerations controlled" (Roper, 1973), Olmsted wrote a pamphlet for publication exposing all the frustrations and threats to one of the most famous urban parks in North America. Prospect Park, another famous park in New York City which Olmsted had a share in designing was experiencing "the usual story: commissioners who understood nothing of the purpose or management of public parks [who] failed not only through ignorance and negligence to realize land-scape effects the designers had had in mind; they also made radical design changes without professional advice" (Roper, 1973). In 1888, the magazine "Garden and Forest" observed that "every park in the country was suffering from the same causes then threatening Prospect Park: irresponsible commissioners and uneducated public intelligence" (Roper, 1973). In the 1880's Olmsted's "life had been embittered and his health damaged by his late experience with Central Park." He told Grover Cleveland "that every public work he had done, and left, was being despoiled by ignorant management" (Roper, 1973).

In 1883, Olmsted, in his sixty's, moved to Brookline, Massachusetts -abandoning his former establishment in New York City, which "had become hateful to him." In Brookline "he carried a heavy burden of professional works, some of staggering complexity" (Roper, 1973).

Green — And Olmsted was not getting along with one of the Commissioners of the Niagara Reservation appointed in 1883—Andrew Haswell Green. Green had accepted Olmsted's recommendations to the State Survey in 1879 only reluctantly (Roper, 1973). Olmsted and Green knew one another. When Olmsted heard that Green was appointed Commissioner in 1883 he "at once wrote Potter and Norton confidentially that it compelled him to 'decline any responsibility that I might otherwise have henceforth to you or others in the matter and to escape my interest in it as fast and as far as I can without provoking inquiries frank answers to which would serve no good purpose" (Olmsted in Roper, 1973).

Green was considered a "public nuisance, an obstructionist and, I almost feel disposed to say, a 'Crank,'" by Howard Potter, president of the Niagara Falls Association. "He thought that Green was coming to understand that people had lost confidence in him and regarded his connection with any public interest as objectionable" (Potter, in Roper, 1973). Green, apparently, "expected to control three votes and the commission," in order to exclude Olmsted's involvement at Niagara. Olmsted indicated that "he would under no circumstances take a paying post under the commission" (Olmsted, in Roper, 1973).

Later in life, however, "Frederick Law Olmsted, who worked with Green on the Commission and frequently disagreed with him, conceded that, 'Green did a hundred times more work than the rest of the commission together'" (New York Preservation Society Archive Project, 2013, http://www.nypap.org/content/andrew-haswell-green).

Commissioners of the Niagara Reservation — Originally there were five Commissioners of the Niagara Reservation. In 1885, sometime after issuing the first annual report of the Niagara Commissioners, Commissioner J. Hampden Robb resigned. Prior to being appointed Commissioner, Robb had experience with the early administration of Central Park. "As a member of the Assembly in 1882, and as State Senator in 1884 and 1885 he was one of the foremost advocates of the law creating the Reservation, and a prime mover in securing appropriations for the purchase of the property." He was a member of the Park Board of New York City in 1887 and became its president in 1888 "his efforts did much to keep Central Park inviolate and to develop that and other parks in New York City in accordance with a wise and consistent policy. He was an incorporator and for the past five years a trustee of the American Scenic and Historic Preservation Society which organization he had aided the late Andrew H. Green to establish" (28 Ann Rep Comm, 1912).

The Commissioners of the Niagara Reservation, however, by 1885, needed a plan by which the destiny of the Reservation would be fixed. "Two members held that the commission's common sense would be an adequate guide to restoring the scenery around the falls and to managing the reservation. Dorsheimer, on the other hand, pressed strongly for a preliminary report from Olmsted, Gardener, and Vaux; and Green, declaring that Olmsted was "particularly offensive' to him, insisted on consulting Vaux alone. The year 1885 passed, and well into 1886 the commission was still muddling about without plan or professional advice and coming under editorial fire for its inaction" (Roper, 1973).

"On October 6, 1886, the commission, unanimously persuaded at last of the need for professional supervision, directed the immediate employment of Olmsted and Vaux to prepare a plan for the state reservation. Green's motion the next month to rescind this action and employ Vaux alone was voted down; and Dorsheimer's motion confirming the previous employment of Olmsted and Vaux was carried four to one" (Roper, 1973).

The body of the plan was presented in the third annual report of the Commissioners of the Niagara Reservation in January 31, 1887, although the plan was written after October 6, 1886 in no more than three months. Olmsted and Vaux visited the Reservation on the 28 of October "and have since been engaged in the preparation of a design for the restoration of the scenery" (Welch, 3 Ann Rep Comm, 1887). The months available for them for work in the field occurred in late autumn, early winter: during November, December and part of January. William Dorsheimer was still president of the Commission, but note that Mr. Green's name is conspicuously absent from the list of Commissioner's names who endorsed the plan to the Legislature in the supplemental section of the third annual report (note, however, that on the plan accompanying the plan for the improvement of the Reservation, 1887, all five of the original Commissioners names are appended).

In the report of the Commissioners, Welch's note of the previous year that "a portion of the land used by the Hydraulic Power and Manufacturing Company had never been acquired from the State" had been investigated and it happened that "the owners of the adjacent lands had never acquired a legal right to occupy the bed of the river and were negotiating to acquire that right" (Welch, in 3 Ann Rep Comm, 1887). The company deeded, then, "part of a lot on Buffalo street" with the restriction that no building be placed on it to interfere with the view. "A part also of the unacquired territory which had been used by them is retained by the State, and will be included in the reservation."

The Commission established a "cheap carriage service" for the island, which facilitated "the rapid movement of the large throngs of excursionists who have come in greatly augmented numbers since the opening of the reservation."

Since most of the offensive buildings had been removed from the Reservation, and their foundations graded, the commissioners were ready to begin the process of restoration of the environment. "The banks and islands, stripped of their natural forests to make way for these buildings now removed, must be replanted."

In the preface to the Olmsted-Vaux plan by the three remaining Commissioners, these men inform the Legislature that the commission approves of the plan "of the eminent landscape architects" Olmsted and Vaux "in its main principles and features." The main principle is "to restore and conserve the natural surroundings of the Falls of Niagara, rather than attempt to add anything thereto." The word "conserve" had not appeared before in the discussions regarding the falls. Whatever grew spontaneously on Goat Island grew by Divine power, not by human ingenuity. This is the fundamental idea of the plan of 1887, and as such follows the plan of 1879.

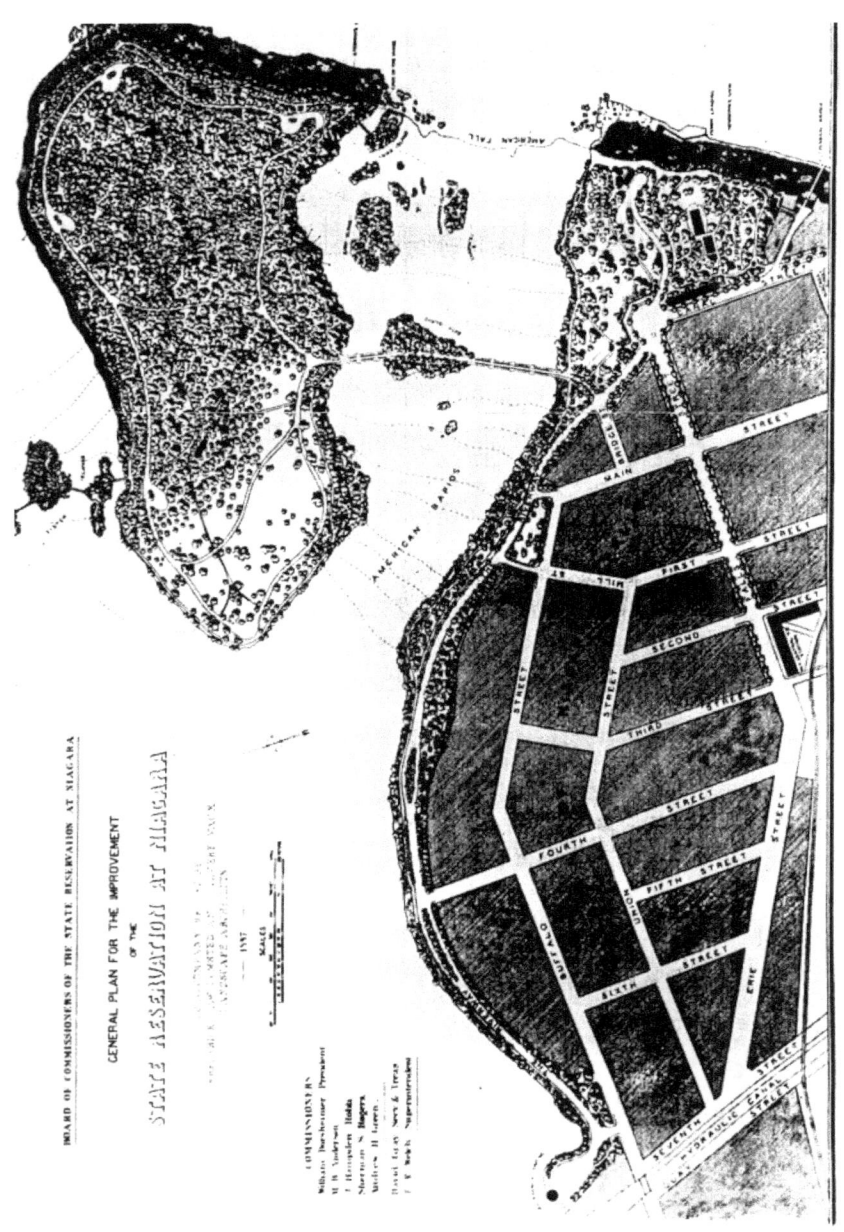

The Plan for the Improvement of the
Niagara Reservation, 1887

The inadequacy of the 1887 plan is that it does not fulfill the promise of the earlier plan of 1879, which so carefully documented the objectives of establishing a reservation. True, the scenery was to be restored, but there is hardly anything of any practical use in the 1887 plan to effect this purpose. It seems to abound in discussions on the magnitude of the coming population, discussions on roads, walks, shelters, to have art or not, balconies, views, iron structures, seats, trellises, bluffs, summer-houses, dams, crib-work, bridges, horses, vehicles, services, cost-rates per passengers, wagons, carriages, throngs, staircases, the order and disorder of visitors and the planting of trees. The plan is so cluttered with these issues that hardly anything clear is proposed for the actual restoration of the scenery. In the years to come, the broad outlines of the structural aspects of the plan would be followed, with technological advances such as in the development of asphalted roads built to suit the speed of the modern automobile on avenues meant for horse-drawn conveyances, but the scenery waited and still waits to be restored while suffering such deterioration over a century of its neglect that little of its 1885 character remains.

One would expect Olmsted and Vaux to have provided details of how to reconstruct the primitive environment on the river—a plan of complex botanical outline. It must be remembered, however, that Olmsted and Vaux were not botanists—they were architects. The ecological principles of plant community structure, composition and interaction basic to plant restoration, were unavailable to these men, and at the time most botanical interest concentrated on taxonomy and phytogeography. Botany had not yet become a profession, but was being pursued by amateurs from other professions such as the medical and legal, and by a class of young, educated women (Rudolph, 1990).

Olmsted's genius was volumetric. He could look out on a landscape and reduce it to three-dimensional masses of soil and vegetation across which were laid harmonious lines of traffic or water in rivers and streams. Visual impact due to the alternation and contrast of open and closed spaces based on an aesthetic interpretation or discipline of the natural volumes disposed in the landscape was a skill he was particularly good at. He could ride to the top of a hill, gaze out, and design an ingenious rural landscape within or adjacent to an urban one (see p. 322 in Roper, 1973, about Delaware Park, Buffalo).

It might appear that, given his particular genius, that individual plants were trivial details, details such as rivals in the field of landscape architecture might enslave themselves to, applying mere flat decoration onto a landscape they could not grasp three dimensionally, that is, spatially, gilding the lily of the land, which was beautiful in itself, a sculpture on which horticulture was as a painted canvas.

Ecological restoration — Olmsted's lack of botanical training, however, he came to view as a handicap which his son, Frederick, Jr., was not intended to be afflicted. "The son had inherited from the father certain traits of mind that made it almost impossible to become highly expert in the scientific classification of plants. Although the field was not completely barred to Frederick's [Jr.] intelligence, it was one he could penetrate only with distressing labor. Like his father, he was not greatly interested in plants as plants; both were alert to their qualities of form, color, and texture and interested in them as elements in landscape composition" (Roper, 1973).

Vaux, too, was made to suffer for his inexact knowledge of botany. When the Tammany politicians of New York City wished to hound him out of his duties in Central Park, "the Commissioners of Accounts of New York City ... had begun an investigation of the park department. Interrogating Vaux, they had baited him—asking him, among other things, the botanical names of flowers he did not know—until he lost his temper and made a show of himself" (Roper, 1973).

As a result of there being no detailed ecological plan for the restoration of the scenery at Niagara Falls, the great promise of all the preceding legislation and activity to protect the falls remains yet only a promise.

There were, however, protocols to follow in the Olmsted-Vaux plan which were and still are very valuable procedures providing some protection for the native plant populations still remaining, and procedures for replanting to achieve a restored environment.

One of the mistaken impressions the plan has made, especially in the present day, is that the specific suggestions Olmsted made were the object of the document, as though carrying them out would create an "Olmsted Park" like Central Park in New York City, or Delaware Park in Buffalo, where a design is imposed on the landscape. At Niagara Falls, as at Yosemite and other areas of natural splendor, the "park" was already there, complete and nearly perfect. Olmsted's suggestions were only to prevent its destruction, maintain and restore it, and allow the public to refresh themselves in its contemplation. If he knew of better ways to protect the unique landscape, doubtless he would have adopted them over any of his own, perhaps even to excluding private vehicles as a means of transportation, construction of boardwalks on the Three Sisters, and other policy changes for protecting State property that might prove superior to those proposed in the 1887 plan.

The primary objective of the Olmsted-Vaux plan was "the preservation of the property of the State" (all quotes are from Olmsted & Vaux, 1887, unless otherwise noted). It was an elaboration of the second mandate (the first was purchase of the property) given by the New York State Legislature to the Commission for the Niagara Reservation once the property had been purchased: "to restore and conserve the natural surroundings of the Falls of Niagara, rather than attempt to add anything thereto." They were to "prevent these provisions from appearing harshly intrusive upon the natural scenery, and to guard the elements of natural scenery from injury and secure their healthy development." Scenery again, as discussed earlier, referring solely to the botanical character of Niagara Falls.

A plan was necessary because such a preserve had no precedent in law or government, and there was real danger that the intent of the legislation would be misunderstood, and that the unrestricted influx of thousands more visitors would destroy the vegetation which had been preserved by the Porters for around half a century. "It is more to be apprehended that waste will come because the main object of the State in making the Reservation shall be lost sight of or become confused in the minds of those engaged in its direction with objects that are wholly foreign to it." That object being the preservation of the native ecosystem.

The Goat Island ecosystem was to remain intact without aesthetic interference by horticulture, architecture and other artificialities or cultural impositions, or destruction for the accommodation of people visiting the island for reasons unsuited to the State's purpose in establishing the Reservation.

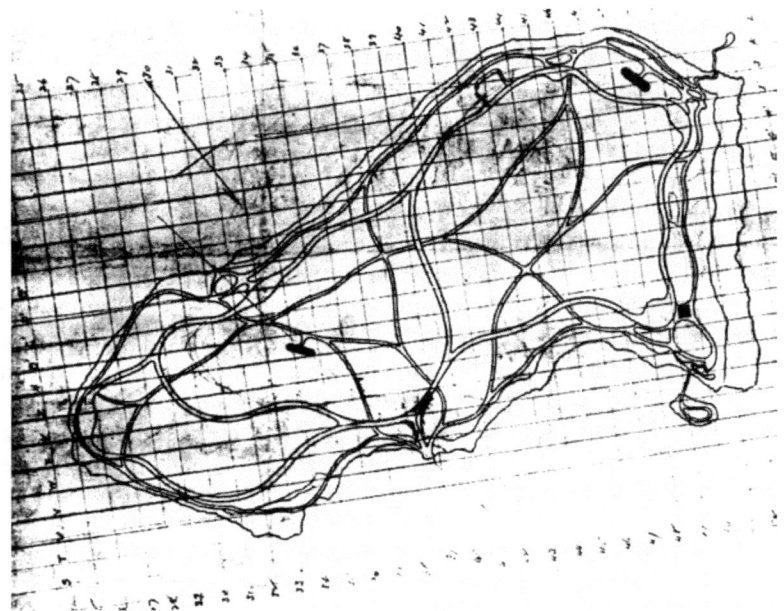

Paths — Following the Central Plan, paths on the extreme margins of Goat Island were 15 feet wide and followed the course of the old carriage road. Important trees would not be sacrificed for the laying the paths. Loop walks would run from the main paths to shaded seats commending views. Paths in the central woods "are designed to be little more than trodden foot-paths and will give forest seclusion to those using them." Strollers could get lost on these paths, so dense was the forest. Instructions and guide-boards posted at the Reservation entrances and hotels, and at various points on the paths would help orient visitors. The reason for such an extensive system of paths was due to a "reckoning upon a turn of custom." For the benefit of walkers would be two large shelters in case of sudden showers. In the woods, they would be "simply large roofs supported at opposite ends of each there should be masonry, without walls, except that at opposite ends of each there should be enclosures for water-closets, and the keeping of police conveniences." Only these two structures and the Cave of the Winds structure, were to occur on the Island. (Plate reproduced from Beveridge, 1985, manuscript division, Library of Congress). Note that the carriage roads proposed were to be lain on new beds, interior to the old carriage road which was constantly subject to loss of sections due to the collapse of the Island margins, particularly on the south side where erosion was most pronounced.

FOREST
SITUATION

RIVER SIDE/
SHORELINE

Seating — In addition to Olmsted and Vaux's slat-wood seats, adaptive use of natural materials for seating is recommended: slabs of limestone/dolomite on the river shore might protect the banks, provide a cool seat, blend in with the natural surroundings, be vandal-resistant and inexpensive, also providing a substrate for the establishment of vegetation. This is especially true of old tree-material carefully placed in wooded or shrubby settings. At the present time, however, most (all) of this material, which could be reused to strengthen river banks, provide "nurseries" for the reestablishment of vegetation, seating possibilities, are presently removed from the Reservation. Note that the placement of large rocks and tree trunks also provides seating for fishing (my drawing).

Again, one conceptual deficiency of the Olmsted and Vaux plan is that what is to be preserved is defined so much in negative terms—that which is not horticultural, not cultural or artificial. The value, philosophy and affects of gardening and beautifying structures is well stated, but not that of aboriginal landscapes, not, for example, boulder communities with their special components of Sedge (*Carex eburnea*), Rock Cress (*Arabis lyrata*), Polypody (*Polypodium vulgare*) and bryophyte (moss and liverwort) species—of *Anomodon, Tortula, Weissia, Fissidens* and so forth. There was no practical or imaginative framework for native habitats for Olmsted and Vaux to use to vie with that already clearly established in the culture with respect to contrived landscapes. The most confusing element in Olmsted's explanation of why ornamentation of the landscape at Niagara is inappropriate is that he uses objects that are clearly beautiful images, such as Japanese embroidery and Florentine mosaics, to contrast with the more desirable, in his opinion, but not generally appreciated natural elements he was attempting to promote. Japanese embroidery, Florentine mosaics and elaborate flower beds may have been more like "beautiful nuisances."

Perhaps, however, the more damaging suggestion, due to its quite suggestiveness, is its application of landscape architectural techniques to the management of the island. It appears that only "decorative" efforts were to be abandoned by the Reservation caretakers. The "operations of road and walk making, the grading of slopes, the dressing of ground surfaces and their clothing with herbage, the planting and nursing of trees, the building of foot-bridges and other constructed objects" will, in result "be precisely the same that it is in ordinary gardening works." Again, Olmsted and Vaux do not make sufficient distinction between a contrived garden and an ecological restoration to prevent continued degradation of the scenery which the Reservation was legislated to protect and restore.

Tourism and planning — At Niagara, Olmsted had witnessed at first hand the "throngs" on the mainland anxious to visit the properties they now owned as citizens of New York State and the Province of Ontario. Now he had a chance to observe what he must have been able to detect as the conflict between public ownership with its attendant public uses and the integrity of native scenery, the "state property" the plan was meant to preserve. Clearly the Niagara Reservation was to be protected as diligently as any of what were to become National Parks— Yellowstone, Yosemite, the Grand Canyon.

"We are assured by the experience of many other places of resort, that, unless extraordinary precautions are taken to prevent it, results will follow of a most unseemly character, destructive of the pleasantness of the place in all respects, and this not only for the time being but permanently, the most secluded and otherwise delightful parts of the grounds, gradually acquiring a disorderly and squalid character, the reverse of that natural sylvan freshness which it is the principal aim of the undertaking to conserve" (Olmsted and Vaux, 1887).

According to the Commissioners, "the plan proposed contemplates the fullest opening of the Reservation to the use and enjoyment of the public compatible with the maintenance, intact, of its natural scenery" (3 Ann Rep Comm, 1887).

The conditions present on Goat Island, in Olmsted and Vaux's experience before this plan was written, were the products of nature and of the restrictions imposed by the judgment and taste of the private owners. Up to that time, the chaos and harassment inflicted on visitors existed on the prospect-areas of the mainland on the American and Canadian shore. Although the public now owned the island group, the Olmsted-Vaux plan was an attempt to apply the beneficial administration of the previous owners, who had managed to preserve the island's wilderness beauty for seventy years, to subsequent public stewardship by the Commissioners and Superintendent of the Reservation.

Access to the falls' environment presented two problems: how to get to the river and gorge margin on the mainland, primarily in the area called Prospect Point, and how to get out to Goat Island and the other islands associated with it. The first situation was controlled by the proprietors who had erected barricades and permitted views after a fee was paid. Visitors to Goat Island had previously been checked by the fifty-cent admission charge and by the effort and time it took to cross over to the island, walk to particular areas of interest, and come back out again. One wonders what condition the roads were on the island before 1885, but it is definite that they were primitive.

Note also the emphasis on building a new road on Goat Island had nothing to do with visitors, but with the fact that the Porter-road was built almost flush with the outer banks of the island. Frequent mention is made here and there throughout the literature of the loss of this road through bank collapses on the south and west ends of the island. The road was to be built in the island's interior (1) for public safety reasons and (2) to avoid the cost of periodically repairing it when the bed was washed or had fallen away.

Local residents, naturally, would be reluctant to visit the river margin and island properties on a regular basis mainly due to the admission fee. While tourists might experience Niagara for its entertainment value, and be glad to pay for the experience, as on entering a theater, local residents were deprived, by the entrance fees, from habitually using the river area for recreational reasons. "One of the most noticeable results immediately following the establishment of the reservation, was the increased number of visitors apparently of limited means. Many of these people lived within a short distance of Niagara, but had been deterred from seeing it by the fees heretofore charged for admission to the grounds" and "At present the expense attending a visit to Niagara Falls is so great as to put it out of the reach of many of those whose right it should be freely to enjoy it" (1 Ann Rep Comm, 1885).

It should be considered that the statistics marshaled in the interests of establishing and maintaining the existence of this public land did not differentiate between visitors to the Prospect Park area on the mainland and visitors to Goat Island: both were "visitors to the Reservation." Some possibility exists that the lists of visitors detailed every year during the term of office of the first Superintendent, Thomas Welch, were derived from their passing the buildings on Green Island, and the cottage at the entrance to Goat Island where the Commissioners and other administrators, such as police personnel, resided. Trains, such as Vanderbilt's New York Central Railroad, continually brought excursionists to Niagara Falls, some walked to the islands, some hired carriages.

By the middle of the nineteenth century some notice had already been made of the problem associated with large numbers of people within the natural setting of Goat Island. "From the sublime pleasures of these scenes, there was but one deduction—the total absence of anything approaching to solitude. The rail-roads of America have rendered Niagara only too accessible: and Goat Island is no longer the wild resort of adventurous lovers of scenery ..." (Gurney, 1841).

The Commissioners in their first report stated that "formerly an unbroken forest covered the river banks and the islands in the stream. The traveler was shut in as it were to contemplate in a grateful solitude the magnificence of the scene before him," (1 Ann Rep Comm, 1885). Solitude and seclusion was as central to Olmsted and Vaux's purpose as leisure was, and if an unbroken forest covered the land in the Reservation originally, then this unbroken forest was to be rebuilt—that is, replanted, so dense and thick as to envelope the visitor.

Seclusion — As I believe I have demonstrated elsewhere (see section on the central woods, Eckel, 2013), conditions were such on Goat Island in 1885 that "seclusion" had a special significance. The woods there was unusually dense in response to the unusual conditions favorable to extensive growth and floristic diversity in the vicinity of the falls. The woods was so dense, rather than open as is typical of beech-maple climax forests where the canopy is usually closed, that Olmsted and Vaux could propose a series of woods trails "through the thick woods," which were to be "designed to be little more than trodden foot-paths and will give forest seclusions to those using them." At Luna Island, so great was the confidence in this enfolding vigor and lushness of the native vegetation, that "bodies of foliage shall here be grown sufficient to secure the larger part of the ground which visitors will be allowed to occupy from the sight of those looking from the superior point of view on Steadmen's [sic] Bluff." This bluff was the high bank of sediments on Goat Island down which visitors had to walk to get to the bridge to Luna Island and famous for the luxuriance of its vegetation, probably due in part to the moisture from mist (see detailed discussion above)—a far cry from the naked lawns on both Stedman's bluff and Luna Island today.

Other recommendations were made consistent with this opportunity for seclusion central to the Goat Island experience. Adjustments to the steps leading down the two bluffs at those parts of Goat Island overlooking Luna Island (Stedman's Bluff) and Terrapin Point (Porter's Bluff) would not only free ascending and descending visitors from confrontation, but separate these two groups of people, possessed of two different sets of emotions, from seeing one another, repeating the ability of the landscape to hide or protect the personal experience of the individual.

The concept of "screening" out the urban landscape which surrounded the islands was not simply erecting a wall of trees, but of duplicating the enclosure of the native woods on Goat Island. Other visitors engaged in similar activities could easily be hidden from each other's view, creating an intensely personal experience, almost an immersion, in the natural landscape found only at the falls.

It had been known from the first that the ordinary visitor to the cataracts spent little time in the area. Even after establishment of the Reservation, in 1893, the Commissioners were to note that it "has been frequently remarked that tourists generally stay but a short time at Niagara. Although the Reservation is annually visited by half a million people, probably ninety-nine hundredths of them spend less than two days at the Falls, while most of them arrive and depart the same day" (9 Ann Rep Comm, 1893). This short duration of stay was to continue up to the present day. Something fundamental to social history or conditions at Niagara has maintained this trend throughout the century.

One explanation for this brevity may be sought in the pace of an increasingly industrialized society which aggressively searches for its weekend and vacation diversions from business with the same efficiency as the organization of the work-place. Although the nation's interstate road systems were originally built for the efficient movement of goods and military materiel, they are also the speedways on which people can drive long distances to get quick vacations further and further from home.

Leisure has a certain refreshing pointlessness and a passivity that contrasts with the aggressive behavior associated with development, expansion, growth and progress characteristic of American society.

Travel — The mechanism by which this historic trend of the quick-visit to Niagara was achieved was the vehicle: the barge on the Erie Canal, the train, the carriage, and ultimately the automobile.

Probably one of the central reasons for the enormous number of visitors coming to the Falls for their brief encounter with nature, other than that there were simply more people in the population generally, and more trains ("by 1853 there were six railroads being built or completed in the Falls area ..., and by 1854 nineteen trains were made up daily in the Falls. A similar number arrived with incoming passengers each day" (Scott & Scott, 1983) was that the force or impact of the experience could be gained in such a short length of time. "Goat Island remained the most popular attraction in the 1860's. Over 100,000 people were predicted to visit it in 1868" (Niagara Falls Gazette, May 13, 1868 in Scott & Scott, 1983). In 1911, Niagara Falls attracted "more than a million people annually, which is at least double the number officially recorded as visiting all the National parks and reservations in the entire country," 28 Ann Rep Comm. 1912. It appeared to be worth the time, money and energy, what little there was of these, to acquire a quick baptism in Niagara's foaming spray on the Maid-of-the-Mist boats, to be struck with an admiring or reverential fear in the high-energy environment at the cataracts, or, as Olmsted and Vaux put it, to be "astonished," and then to go away.

In the face of these facts, which were apparent before the Reservation was established, Olmsted and Vaux, with the Commissioners, sought to slow the natural pace of the "throngs" which came to Niagara by forcing people out of their carriages to take longer time to enjoy Niagara. The Porters had not improved their roads and for three quarters of a century or so the public had not complained, and the environment had flourished. Naturally, to continue to preserve the scenery of that environment, Olmsted and Vaux tried to preserve the pace of visiting it. "Realizing that the majority of visitors to Niagara will always explore its beauties on foot, the landscape architects have given the preference to this class, over those in vehicles, at all points of view where space is limited" (3 Ann Rep Comm, 1887). The desire to downplay vehicular traffic on Goat Island was repeated by the Commissioners later: "Indeed, it is impossible, except on foot, thoroughly, to see the principal charms of the Niagara scenery" (6 Ann Rep Comm, 1890), although it is probable that most visitors, especially those whose homes are several days journey from the falls, do not explore the Reservation in this way. As recently as 1982, the present state park agency stated "there is no question that the pedestrian can gain greater enjoyment of the Reservation ... [the agency] is also aware that most park patrons do not take the time, or are physically unable to undertake a walking tour of the park" (state spokesperson in Otis, 1982).

Olmsted recognized a very important element in his treatment of the number of visitors to Goat Island: these anticipated "throngs" would not swarm indiscriminately all over the island, but would concentrate only at certain vantage points—hence there could be seen, as early as 1886, that there were two kinds of visitors: those who came for the brief contact, and those who stayed. Design of Goat Island would reflect the needs of both kinds of visitor. It is probable that the local residents (living within one day's journey of the falls) would ramble, but those coming from farther away would be quick about it. The latter, too, would tend to concentrate on the Reservation in the summer (vacation) months, the former would come generally throughout the year, during their lunch hour, on the week-end. Many of these would be the labor force of the City of Niagara Falls. This characteristic of the visiting population recognized in the Olmsted and Vaux plan is much as reported today (see "Patterns of use/tourists versus local users, in Promontory Partnership, 1981).

"... Great throngs of visitors are to be anticipated only during a short period of the year, and then only during certain hours of the day. During these few hours those composing them will rarely scatter far from certain well defined localities and the routes of movement between them. They will but little occupy parts of the Reservation most delightful to ramblers ... few visitors [intending to spend no more than two days at Niagara] ... will have time ... for the quiet strolling and resting, through which alone the more secluded beauty of the Reservation is to be contemplatively enjoyed." The focus of Olmsted's suggested alterations in the flora was restricted to certain areas of high use. The rest of the island would be left alone.

An important concept Olmsted and Vaux had was to distribute the numbers of people throughout both the Canadian and American reservations. They were to keep separate those in vehicles and those on foot. The automobile, which neither author anticipated, would introduce a third element due to its higher speed, requiring more space to operate safely, that is, visibility would have to be expanded and consequently trees and even land-forms eliminated. Standard automobile-traffic safety factors such as removing trees to create lawn verges on either side of the roads so cars accidentally leaving the roads would not crash into tree trunks is another example of the inappropriateness of the personal automobile on Goat Island—an extension of the problems of the horse-drawn carriage in the 1880's.

In the present day, in the interests of forcing the attendance numbers higher than they would normally be, however, those areas on Goat Island which do not experience high attendance were considered "unused" with the result that the ecosystems there are disrupted when these areas are given over to picnickers, car sales, and helicopter rides (see recommendations, the Promontory Partnership, 1981). It is exactly the "unused" portion of the Goat Island woods that was chosen for the big parking lot built in 1951.

Protecting the ecosystem — Hidden in the inadequacies for vehicular amenities in the Olmsted-Vaux plan was the desire not to sacrifice the ecosystem for the sake of the logistics for dealing with the conveyances in which people arrived at the Reservation. The interesting point is that it was understood, but inadequately acted upon, that the unassailed vegetation would have to go if there was to be vehicular traffic on the island. Even though many more people were expected to demand to see Niagara from the vantages of Goat Island, and from their vehicles, Olmsted trusted that subsequent managers of the Reservation, because their mandate was, in his view, clear-cut, would always manage the public in the interests of scrupulously maintaining the vegetation. As will be discussed below, subsequent history has shown that the reverse has been the case.

Olmsted proposed a very narrow road: around twenty feet wide, and in places perhaps narrower still "in accommodation to the trees between which it is to pass." A normal road was twice this, and on a city street, up to three times wider. They were criticized for this idea, but they justified it by the number of lay-bys to be constructed, the single direction of traffic, and the proposal of two forest roads bisecting the central woods for those wishing not to make the circuit of the island. Only one central road later came to be established. Todd (1982) reported 10,000 visitors to Niagara Falls per day "even in the 1880's," a fact which lead to the "exclusion of carriages on the grounds of efficiency" (Todd, 1982).

The plan to slow the tourist population down rankled those whose business was transportation. In the wrangle with the New York Central and Hudson River Railroad, the correspondence of which is included in the seventh annual report (1891), an agent of that company wrote to the Superintendent of the Reservation that:

"It is not questioned that the idea, that Niagara can be appreciated in a rapid ride about, is absurd. Days, of course, should be passed about it to worthily comprehend its grandeur and beauty. But the fact remains that such is not the popular idea, and until that idea has been removed by educational influences, not at present within our control, our movements must be regulated with the idea of protecting people who imagine they can comprehend or even see Niagara in a day."

Olmsted knew the dangers to protection of the vegetation that the vehicular infrastructure on Goat Island would become. His plans for the restriction of vehicles on the island was apparently made with the broader understanding that the Canadian park, made in conjunction with the American one, would better bear the burden of views taken from vehicles. On Goat Island, only the limited breadth of Terrapin Point and Stedman's Bluff were available for prospects onto the cataracts, whereas the Canadians commanded The Front: a one-mile unimpeded view of the entire cataract system. "... when the improvements proposed in the plan shall have been fully carried out, and when, to these, the improvements to be made on the Canadian Reservation shall have been added, the number of persons at any time visiting Niagara, will be much more extensively distributed, and will be much more in circulation ..." This was in reference to Olmsted's consultation with the Commissioners of the Queen Victoria Niagara Falls Park. "The preservation of Goat Island, in particular, depended on the availability of good carriage views of the Falls from the Canadian reservation. Visitors must be willing to take their carriages there instead of thronging to Goat Island and the edge of Prospect Point" (Beveridge, 1985). Apparently to this effect, the Canadian administration established its road closer to the gorge crest than it had originally intended (Beveridge, 1985).

Note that in addition to parkage areas for vistas at Stedman and Porter's Bluffs, a parkage was established by the Spring as apparently this area still was the sole source of water for refreshment on the island.

Traffic — The reason why the plan failed to limit a destructive traffic system in the Reservation was not because the plan did not accommodate the people, but perhaps because the plan compromised where, with the hindsight of a century, it should not have. No more carriage traffic should have been allowed on Goat Island than had been allowed for the previous seventy years. Olmsted knew "extensive pleasure driving" would destroy the "sequestered woodland beauty" of Goat Island. What the plan amounted to by the improvement of the road system was an ambiguous agreement that visits in vehicles to the island would be permitted, and modifications made to the forest for the sake of roads and all the associated changes of drainage, grading, visibility and right of way, the necessity for parking and turn-arounds, of paving and repaving and curbing and salting the road in winter, altering the chemistry of the soil. Olmsted and Vaux thought they could control this direction by proposing a road-system so inefficient that people would be forced to descend from their vehicles or perhaps abandon them on the mainland. Naturally this would have bothered visitors as much as the muddy roads, and later these inefficiencies would be corrected, with the subsequent loss of the natural conditions these men wished so strongly to preserve, the most extensive being the sacrifice of seven thousand square yards of the richest section of the Goat Island woods for a parking lot, begun in 1951 (Scott & Scott, 1983).

The context, however, in which Olmsted and Vaux couch their proposal is the establishment of a means of public transport for the island, not private. They described the establishment of a "route-carriage system ... the methods of use and manner of payments of which vary not essentially from those of the omnibus and street-car system of large cities." In the interests of such a service, the roads are to be improved and the cost rate per passenger reduced "by a finely economical adjustment of horses, vehicles, movements and stoppages to the peculiarities of the service ..." They hoped fondly that "after a few years but a small proportion of the visitors to Goat Island will use any other form of wheel conveyance, and the road system of the plan has been devised accordingly." Had Olmsted and Vaux known of the coming independence of personal movement to be brought by the automobile, they may have put restrictions on private conveyances to the island, and prohibited the sacrifice of the primeval forest for the convenience of dropping off ones vehicle near the falls to take a look at them.

When the Reservation administration did provide its own conveyance system, it was met with approval by the distinguished geologist, Amadeus Grabeau, then at the Buffalo Museum of Science, who agreed with Olmsted and other advocates that Goat Island must not be seen from a carriage, but "if you must ride, patronize the reservation carriages, which leave you wherever you wish to stop and take you on again at your own pleasure." And in a footnote: "These carriages are run at intervals of 15 minutes, starting from Prospect park, and making the circuit of Goat

Island. The fare is 15c for the round trip, and stop-overs at all places, and for any length of time on the same day, are allowed" (Grabeau, 1901).

The inadequacy of the 1887 plan in the century ahead was also due to the encouragement of quick visits by subsequent government administrations, because of the political value of attendance numbers, in addition to the generation of revenue by charging a fee for parking—the latter a form of commercialization of the Reservation. These factors play a major role in justifying an infrastructure in the Reservation favoring access by personal vehicle, regardless of the ecological cost.

Olmsted clearly saw the tendency to increase attendance at the expense of the primeval habitats on Goat Island. Originally, high attendance rates and quick circulation of tourists (turnover) were for the economic gain of private interests—later the rates became manipulated for political and economic interests, such as to promote the local and state tourist industries. Olmsted and Vaux stated clearly that the industrial buildings be removed and "many things originally regarded as luxuries for the entertainment of visitors, especially of the great illuminating apparatus; in preventing the approach of a railway for the accommodation of visitors, because of the injury to the scenery that it would entail, and in forbidding exhibitions in or over the waters of the Reservation, *the effect of which would be to attract a larger number of visitors to it for other reasons than those presented in its natural scenery*" (my emphasis).

The belief that people would be satisfied with a restricted private-vehicular access to the Goat Island complex if such access was allowed at all is repeated in a recent proposal recommending road changes there (The Promontory Partnership, 1981)—primarily by reduction of vehicular parking to below capacity during peak visitor periods and other improvements in the road systems on the adjacent mainland to relieve vehicular congestion on the island.

It is apparent, given the historical perspective of two centuries, that the vehicle (automobile) and the wilderness state of the Goat Island property are incompatible.

The State itself has to provide motorized access to the islands if the area is to be returned to the wilderness setting which the State promised its citizens it would do. Alternative modes of access would be those accessible to the public for two centuries: by foot, by bicycle and mass transit vehicles, such as viewmobiles. Reestablishing a horse and buggy service on the Island today, such as is currently available in Central Park, might be an attractive choice for transportation, returning part of the maintenance buildings located there, perhaps inappropriately, to their original function as a horse barn.

Trampling and other pressures — Olmsted makes very clear again, the disintegration of the natural plant communities with increased attendance, with the implication that this could worsen in the future. He cites the Three Sisters and Luna Islands, which in fact today have had their floras stripped and their soils sterilized by the trampling of visitors. The vegetation there originally appears have to been dominated visually or conspicuously by their bryophyte (moss and liverwort) communities, probably owing to the abundance of moisture and shaded boulder tops these organisms favor. In the 1830's the Three Sister Islands were called the Moss Islands (Scott and Scott, 1983). Olmsted reports that in 1887 at the Three Sisters "a large part of the old ground verdure ... had been killed out, and mud or dust was often found in its place," exactly the condition which exists today only far more advanced than what Olmsted witnessed. The rock vegetation had been stripped which once was "remarkably luxuriant, low massive bodies of a description of foliage rare even to most horticultural visitors ..." which included Ground Hemlock *Taxus canadensis*) "our native yew, a shrub supplying the darkest green and the brightest red of our forest." He described the exposure of root systems where paths crossed them and banks were collapsing. Olmsted saw that trampling would become a serious threat to meeting the objectives of the legislation creating the Reservation: "I have seen a great vigorous oak tree killed in two years by the trampling of the ground over its roots ... No turf in our summer climate will remain in any spot where a hundred footsteps have fallen in rapid succession" (Beveridge, 1985). Soil compacted to the density of rock is impervious to seeds becoming naturally established and remain bare year after year—as can be seen vividly at the First Sister Island and adjacent Goat Island shore.

It is difficult to imagine how trampling could be restrained without elevating people above the vegetation being trampled to death by their movements. Although by exactly what structure would be appropriate, Olmsted and Vaux did not venture to say, perhaps because such devices had never been designed before: that "many parts of the reservation shall be prevented from being gradually made desolate by like process [of trampling], an extent of artificial accommodation, and of artificial expedients for protecting nature, must be provided that would otherwise deserve condemnation."

Perhaps, as in the case with many of our nature preserves, structures such as boardwalks (see below for suggestions for a boardwalk system on Goat Island) were implied. Unfortunately, Olmsted did not present a remedy for the process that has nearly destroyed the vegetation of all the peripheral islands connected by bridges in the Goat Island complex. It is hoped that he didn't intend a gate and turnstile system to be enacted, as suggested to control the throngs by way of visitors being "constrained to courses through which any important elements of scenery or any property of the State shall unduly come to injury."

At certain points of interesting view, Olmsted and Vaux intended to provide hidden resting areas. Such areas would have seats of "slat-work, darkly stained and at points fortified with metal, the object being to reduce to a minimum the opportunities for penciling and cutting them so irresistible to a certain class." Some would have "simple trellises over them upon which canopies of vines and creepers, natural to the region, are to be trained." In addition to these seats could have been placed logs from fallen trees and stones or rocks from elsewhere in the region of a natural appearance on which visitors could sit and admire vistas. These would be the most natural to the habits of people, such as is evident in old photographs of visitors during the Porter ownership, of no maintenance requirements, of an absolutely unobtrusive quality, and providing additional substrates for the establishment of mosses, liverworts, lichens, and other organisms. People could attack them to their heart's content. This activity would only make it easier for the creatures named to establish themselves.

"The Reservation includes a part of a bluff by which the riverway abreast the upper rapids is bent toward the river. Upon the face of this bluff there are some good trees, and from the upper part of it there is a fine view of and over the placid water of the river above the rapids." This bluff, or part of it, is still in place, together with its "good trees" of Oak and Maple and other, more recently established species. It is isolated from public enjoyment by the Parkway and vehicular bridge to Goat Island. The herbaceous layer in this small woods is being overrun with weedy, noxious plant species due to the present maintenance technique of clearing out the native underbrush.

One of Olmsted and Vaux's principal recommendations was "that the taking of provisions of any kind on any other part of the Reservation [other than to an area on the mainland portion of the Reservation] shall be forbidden by ordinance and really prevented as a cardinal necessity of the success of the plan." Indeed, it is the ignoring of this detail that has permitted a policy involving degradation and removal of the native flora to provide extensive picnic areas in the wooded areas of Goat Island ("the establishing of a large picnic grove in the wooded portion of Goat Island" Conservation Dept., 18 Ann Rep for 1928, 1929), the building a cafeteria with within a prominent view of one of the most significant wooded areas on Goat Island in 1967 (Scott and Scott, 1983), and maintaining, in the 1980's, an overextended road and private vehicle infrastructure on the island in order to service the cafeteria (Otis, Final Environmental Impact Statement, 1982). As a matter of fact, motivation to install the large west end parking lot may have been to service this dining facility, and thereby demonstrate the spiraling relationship of interacting facilities predicted by Olmsted and Vaux.

Services — In the 1980's conflict between park and the economically depressed adjacent community has arisen because of the park administration's decision to duplicate municipal services within the Reservation—to the detriment of the natural environment (note within the past decade, in 1986, the removal of sixty trees in front of the restaurant on Goat Island to promote the view for customers partaking of restaurant fare, and the public outcry in "letters to the editor" against this decision).

There are two extended dining areas now on the island's west end, and two souvenir shops adjoining them—in addition to a dining area recently established at Prospect Point. Olmsted and Vaux's belief that public facilities for eating on Goat Island would lead to a spiraling of associated developments detrimental to the vegetation is further validated by suggestions recently made to the State that the Terrapin Point area be a staging ground for "seizing the opportunity to utilize one of the most breathtaking natural prosceniums in the world" by "projection on the mist" using a "high intensity light source system atop a structure associated with or close to the restaurant building, the reflected foot lamberts on the mist would provide highly visible images for a magnificent sound and light show" (The Promontory Partnership, 1981) with the potential for additional music and dance, art and craft activities. Indeed, it is because of the present restaurant at Terrapin Point, "cultural events proposed for the [natural] amphitheater above Terrapin Point" are focused. Conveniences, "an office of advice and guidance, a check-room, and large lavatory, toilet and other conveniences" were to be located, according to Olmsted and Vaux, in an area fairly close to where the Rainbow Mall of the City of Niagara Falls is presently situated. Picnicking also was to be located there: "conveniences shall be supplied for those who wish to eat provisions that they have brought with them" including large shelters where people could get in out of the rain. Olmsted and Vaux stated that the influx of visitors would "tarry" or linger in this area before wandering in amidst the various natural features to be experienced.

The administration building would be located in this area, on the outer margin of the natural areas protected by law: "a superintendent's office, with, nearby it, storage rooms, tool rooms, and workshops for repairs. These should be accessible by carts, without entering the roads of the Reservation, and should not attract the attention of visitors within the Reservation." The administration was to maintain its presence in the old building on Green Island and in the maintenance facility at the entrance to Goat Island both built and used under the Porters up until the event of Superintendent Welch's death. Indeed, Welch had removed these structures along with over 150 others from the Niagara Reservation (Scott and Scott, 1983).

Contrary to the plan to preserve and restore the Reservation's vegetation, and contrary to the efforts and intent of Superintendent Welch, who had died in 1903, in 1912 "grading started at the ... gravel pit near the Spring to prepare the area for a new labor center" (Scott and Scott, 1983). By 1915 the "Labor Center on Goat Island was finished to include one building with a workshop, foreman's office, and stock room, and a second building with a dining room for Reservation employees, lavatories, and a coat room. A rubble stone barn was also finished in this work area" (Scott and Scott, 1983). The only structures to be had on the islands were intended solely to benefit visitors, and these were to be unobtrusive by vegetative screens: either trees or viney growth—all administrative operations were to be located out of the central botanical areas. Only seats and two large shelters for the protection from sudden showers, "simply large roofs supported upon piers of rough masonry, without walls, except that at opposite ends of each there should be enclosures for water-closets and the keeping of police conveniences. They are to be the only things on the island of the character of buildings" except the Cave of the Winds shelter. Later, an extensive administrative complex constructed on the island would lead the way to the development of other structures contrary to the Reservation's intent and the plan for its development.

SHELTER ON GOAT ISLAND.

GROUND PLAN

Shelter, details of plans — The plans for the Goat Island shelter at the end of the bridge to Bath Island were published in the eleventh annual report (1895) for the year 1894. They were submitted and approved earlier, as reported in the tenth annual report (1894) for 1893.

Shelter, as built — It is apparent that this shelter, built on Goat Island, was designed by the firm of Marshall L. Emery and Downing Vaux. These mean were paid $176.36, under "shelter building," and Emery $74.25 and Vaux $51.75 for traveling expenses in 1895. Vaux and Emery were also paid $146.48l under the item "terminal station" for the same year. William Shepard & Son performed the masonry work on the shelter building. Note that in this year Samuel Parsons, Jr. had come up from New York to advise in horticultural matters, perhaps with Theodore Wirth. The former is entitled "landscape architect," the latter "landscape gardener."

"The shelter building is situated on the river bank among the trees, on Goat Island, near the bridge leading from Bath Island. It is an inconspicuous structure; and the graceful lines and warm color of the roof, the red-brown tone of the stonework, harmonize with the natural surroundings. It is very readily accessible, and the accommodations for visitors are ample. The walls are of Medina stone, with arched openings; the posts, of Georgia pine, have curved brackets, supporting the over-hanging roof. The floor is asphalt, and the ceiling is of narrow Georgia pine and semi-circular in shape. The lavatories are lined with brick, with slate fittings, and are thoroughly ventilated.

"Permanent wooden seats are attached to the walls, both inside and outside of the building.

"The total costs of the Shelter building was $6,572.94" (13 Ann Rep Comm 1897, for 1896).

Vaux fence — This picture shows the inward-leaning fence designed by Calvert Vaux. The fence was to provide minimal visual obstruction or distraction, and yet provide security against accidents at the brinks of the river and the gorge bluffs (15 Ann Rep Comm, 1899).

Botanical Restoration

Custodianship of the primeval ecosystem was not to be completely passive at first. Eroded banks were to be restored to their original communities, naturally regenerating growth was to be encouraged and disease and natural destruction were to some extent to be corrected to promote the life of the plant, tree or shrub. Later, these suggestions this would come back with disastrous force when the Superintendents of the Reservation would stage great pruning efforts on vegetation damaged in the naturally violent environment of the falls, and where the central woods was to become gutted (thinned) to "promote growth" on the one hand, and "reduce growth" to encourage visitors leaving designated paths, on the other. A lack of the appreciation of destruction and decay in the healthy natural processes of ecosystem self-maintenance was and is not appreciated. Continual maintenance-disturbance, as in the case of continually mowing lawnscapes which inhibits the natural reestablishment of vegetation and promotes the establishment of significant weed populations, later became the norm.

In addition to the ecological restoration, botanical restoration must include attention to the many rare and endangered species known for the Reservation (Eckel, 2013). The immediate vicinity of Niagara Falls is well known for western New York State as one of three centers for rare species protected by law, and is the major station for limestone-loving species also protected (Zander, 1976).

Because the authors of the plan intended some sort of ecosystem restructuring, but had no technical background to assist them, ecological improbabilities are frequent in the plan. They appear to have been conceptually burdened with formulae for planning the multitude of urban and private landscapes characteristic of their businesses, or the plan was written in such haste, and under a variety of pressures, that some of the contradictions between their objectives and their suggestions for improvements did not occur to them.

Such conceptual vulnerability in the Olmsted-Vaux plan which would contribute to the frustration of their objective to protect and restore the native scenery in the decades ahead, was their apparent compromise or confusion regarding natural regeneration versus horticultural or sylvicultural modification. There are frequent references in the Olmsted and Vaux plan to ambiguous positions on protecting and restoring the natural plant populations. The protection part was easy—simply do not disturb what is already there. The restoration part appears oddly artificial in places, mainly because there were no specific references to species or any details of species relationships (communities). Modifications were to be made in the spontaneously regenerating plant communities at certain points, for example, to provide vistas between heavy tree canopies and trailing vines, as given below with respect to the map-diagram that accompanied the 1887 plan (for additional details on relating the plan to specific areas in the Goat Island complex, see botanical descriptions of these areas in earlier sections).

The plan is very positive where it stipulates interactions between natural regrowth and artificial changes, such as using natural regrowth to hide recent constructions. On several occasions Olmsted and Vaux recognized the superiority of naturally established specimens of native flora; as for the "drive on the mainland" they recommend it be "at points divided in order to avoid injury to a few promising trees of spontaneous growth." Its weakness lay in certain of its recommendations which are contrary to the inherent dynamics of the ecosystem and reflect the simplistic notions of vegetational restoration and species preservation characteristic of the times.

Not providing clear demarcation between what was truly natural and what artificial, Olmsted and Vaux indicated that regionally native plants were to be established in areas in seeming imitation of the natural character of the primeval plant communities in place on Goat Island—not without justification as there was no science of ecology as a frame of reference. They recommended planting indigenous species of river-side plants such as "dwarf willows, rushes, ferns, irises, flags" on the proposed short walls built along the mainland parts of the reservation "with a result in view that would differ but little in character from that of the natural, low, rocky shores of the neighboring islands." No specific provision was made to study exactly what constituted the native communities here. Indeed, the plan was drawn up after the growing season was effectively over, and the trees were beginning to lose their leaves—in late October. The plan and map were devised during the late-autumn and early winter months of 1879.

Exact restoration not attempted — Indeed, "an exact restoration of the old shore" with its "gentle undulations toward the river, the immediate margin of which was in some cases flat and boggy ... with a surface partly strewn with boulders and overgrown with bushes and grass" was "not to be attempted." Again, ambiguously, "only its original character is intended in the plan to be regained ..." after the structures were removed. It is ambiguous whether Olmsted and Vaux indicated it was a topographic restoration which should not be attempted, or a botanical one, but as the passage continues, once a topographically acceptable (safe, erosionless) shoreline was established, then plants would be set to match nearby communities on the islands.

If the native ecosystem was to be preserved, why would Olmsted and Vaux characterize its original character, as "overgrown" with bushes and grass? The whole of the primitive areas of Goat Island, which were so valuable, were "overgrown." Why deny the shore the boulders which occurred there originally? If one had gone to the shores of Goat Island at the time, one could have easily seen boulders covered with a rich and characteristic verdure -especially in summer. Why deny this in the shoreline restoration? We also know today that "flat and boggy" river edges can support rare and interesting plant communities, and do. Just a mile or so upriver in the Little Niagara River separating Cayuga Island from the mainland exists a tiny shoreline plant community dense with wildflowers of all kinds, some conspicuously beautiful and rare in the region (for example, Water Willow (*Dianthera americana*). When Olmsted was designing the lagoons for the Chicago Fair in 1892, during a trip to England, he "found a fertile source of inspiration in the vegetation along the Thames banks, especially the willows. Their variety of age and size and color, their intermingling with sedge and rushes and other common native water plants, their different growing positions—sometimes horizontal where a bank undercut by the current had dropped over, sometimes growing in water, sometimes overhanging reedy plants growing on water-covered shoals in front of them—all were suggestive of various ways the lagoons at the fair might be treated" (Roper, 1973).

Prospect Point, 1875 — Cliff plant communities lost through indiscriminate dynamiting of the gorge face in the vicinity of the Falls (both from the 8th Ann Rep Comm, 1892). "Nor have I found anywhere else such tender effects of foliage as were once to be seen in the drapery hanging down the wall of rock on the American shore below the fall, and rolling up the slope below it, or with that still to be seen in a favorable season and under favorable lights, on the Canadian steeps and crags between the falls and the ferry, … the exceeding loveliness of the rock foliage … I believe to be a direct effect of the falls. ... Something of the beauty of the hanging foliage is also probably due to the fact that the effect of the frozen spray upon it is equivalent to the horticultural process of 'shortening in' compelling a denser and closer growth than is, under other circumstances, natural." (Olmsted in Gardner, 1880).

Faking nature — It is not difficult to see how subsequent administrations, without informed biological guidance, would maintain the mainland section of the Reservation as simply a grassy verge with scattered trees of little natural botanical character or interest.

The easy extension of this idea would, in the years to come be to plant to achieve a "natural effect," that is, to "fake" nature. In one recent proposal, it was recommended to develop on Goat Island a "naturalized zone" including non-native species of trees which would "comprise a plant palette similar in form, scale, and texture to the indigenous plants" (Promontory Partnership, 1981), in other words, pretending to a natural plant community, rather than reestablishing genuine native plant communities. As the years of the Reservation wore on, if native stock could not be purchased, or if transferred and it failed, then horticultural varieties were acquired and these were accepted as long as they maintained a "natural character." Welch, in his first report (2 Ann Rep Comm, 1886), remarked on the shrubs found on Goat Island that "many are very rare, and cannot be obtained at American nurseries," highlighting the probability that, to some extent, stock for re-vegetating the mainland and Bath Island had to come from culling the woods in Goat Island—a dangerous precedent leading, over the ensuing century to culling the woods for a variety of reasons until it retains little resemblance today to its precious original character in 1885.

The trees at Prospect Point "are dilapidated through the effects of the freezing of the spray from the falls upon them. They are more or less rotten-hearted and not to be depended upon; nor can any trees be expected to be grown with lasting good effect in the locality [therefore] it cannot be hoped that a pleasing natural character can be fully attained throughout the district." This particular area had been a prospect area for centuries (at least a century before a bridge had been made to Goat Island). Its natural ecological characteristics had been long forgotten, much as is the case all these areas today. Those characteristics were probably no different from those at Terrapin Point (Porter's Bluff) and Stedman's Bluff above Luna Island—in the case of the former, even down to the dolomite flats. Natural plant communities had originally developed with native species compositions giving the community a structure capable of flourishing in the locality named. Artificially maintained horticultural plantings with no sensitivity to the special conditions in the spray zone at Prospect Point and Park were what Olmsted was referring to. Similar complaints were not made a century ago in the still relatively intact Terrapin Point area, which possessed the same physical characteristics, but identical complaints exist now, for identical reasons—the diverse natural communities with adaptable species have been abolished and are unremembered.

Abandonment — The abandonment of a restoration objective in Prospect Park by Olmsted and Vaux creates an unfortunate precedent for the abandonment of a similar objective at Terrapin Point today, presently the site of a seriously degraded natural landscape (that is, reduced to a lawnscape), both inherently due to a lack of investigation into and under-standing of the environmental conditions in both areas, and the develop-ment of a restoration protocol sensitive to the biological needs of vegeta-tion in these areas.

The Porters maintained Prospect Park until it was sold in the 1870's (Porter, 1900). Since they planted much Norway Maple and other exotic trees on Goat Island in the area of the Spring, it is probable that they had modified the native species composition on Prospect Park as well, and Olmsted's observations only reflected the inappropriateness of "urbanizing" prospect areas receiving winter mist. Olmsted and Vaux perhaps despaired of reforesting Prospect Park due to an inability, in such short notice, to observe the real reasons for the conditions under observation.

Another example of an inappropriate cultural policy, rather than an appropriate biological one, is Olmsted and Vaux's suggestion that the denuded, vegetationless "west," actually the south, bank of Goat Island be clothed with "foliage and verdure," but rather than investigating what might be the normal patterns of re-establishment of native species, they recommended trees be planted along the top of the bank to shade the walkers there, and to plant nothing below the bank that would "grow above the line of sight, toward the Rapids, of visitors standing on the walk that is laid out along the top of the bank." If the trees planted along the top shed their seeds, naturally they would germinate and rise to screen the view. The only way to prevent this was to plant alien species of trees whose seed cannot germinate in the physical condition of the islands, or institute extensive maintenance activities to prevent arboreal native vegetation from establishing itself along the shore.

Note that, as discussed in previous sections, this south bank was completely deprived of vegetation due to natural slumping of the soil bank there into the river.

Reestablishment — As a matter of fact, many fine specimens of native trees have become reestablished on this slope and may present an example of natural regeneration using native tree species. Whether these trees, some probably close to a century old, grew by natural or deliberate establishment is not presently known. They included native Basswood (*Tilia americana*), White Ash (*Fraxinus americana*), Sugar Maple (*Acer saccharum*), Hop-Hornbeam (*Ostrya virginiana*) and Yellow Oak (*Quercus prinoides* var. *acuminata*).

A line of trees along the walk presently at the top of this bluff on the south and west sides of the island had been planted quite some time ago perhaps in accordance with the 1887 plan. Native trees were not used, but instead, Black Locust (*Robinia pseudacacia*). Although not typically escaping, numerous saplings of this tree have been found on the lower slopes and flats on the south side of the island, indicating this alien weedy tree is spreading in the complex.

The barren mainland, when recently divested of buildings all along it, was to be the subject of reforestation. This was to be done quickly with good, undiseased stock from the forest on Goat Island. This, together with Bath Island, was the "seventh part" of the area of the Reservation which presented "an objectionable artificial character, most of it for example, having been heaped up or dug out in connection with road or building operations." The Olmsted-Vaux plan looked "to operations the aim of which is to re-establish a permanently agreeable natural character, harmonious with that of the undisturbed parts."

Harmonious, but somehow not identical. Their treatment of the mainland restoration shows specific constraints on the natural dynamics of forest regeneration. This denuded area was to be "ultimately covered with forest trees," which were to be of the same species as grew on Goat Island, but not to be derived from the Goat Island forest. The first Superintendent, Welch, however, did disrupt the native woodland communities there, using that material as stock to plant open areas. The recommended species, in effect Hop-hornbeam, native Basswood, Sugar Maple, Red, Scarlet, White or Yellow Oaks, White Pine, Hemlock, Butternut, Walnut, Hickory, etc., were to be "planted thickly" and thinned of damaged, unhealthy trees. They were to be so thick that their boughs would interlock. There was to be no sylvicultural attention to regimented tree-placement or tree-form, much as indicated by the illustration in the 1879 report. "Individual tree beauty is to be little regarded, but all consideration given to beauty and effectiveness of groups, passages and masses of foliage."

Reference is made several times to the inherent thickness of regenerating growth, and the application of "thinning" techniques, such as to the mainland plantings, as they grow. Only the most vigorous trees would be left standing, ones able to tolerate natural conditions at Niagara. The fact that thick regrowth, that is, structural density of the woods facing the river, the falls and the prevailing winds, might have been critical to the healthy condition of the native woods at Niagara—a characteristic previously noted by many botanists visiting the area—apparently did not occur to Olmsted and Vaux.

A great diversity of trees was to be planted, lest one species succumb and thereby create a visible interruption in the forest cover. The growing trees would be periodically thinned, rather than letting the less vigorous die out, anticipating that only twenty five percent of the established trees would ultimately remain. The whole forest would be vigorously managed, even though not in regimented patterns, such as in an orchard, but still with "centre lines of which spaces will be diagonal to the shore line, in the direction that will leave the Rapids open to view from the drive." Underbrush, "native ... of the neighborhood" is to be introduced only in intervals and only enough "to prevent ... a grove or orchard-like monotony of trunks."

It would be difficult to interpret these instructions as "restoring" a primitive condition on the mainland when there were to be so many artificialities imposed on its character. For example, these trees were not to be arranged as in an orchard when in fact they were so arranged—but this was only apparent when one was looking upriver, perhaps at an angle of 45 degrees to the line of the shore, not across at an angle of 90. Was this to be a real forest, or an orchard, carefully maintained so as not to appear monotonous or artificial, but profoundly artificial nonetheless?

Shrubs were to be heavily planted on the river margin, less so on the upper slopes, primarily because they are short and would not so much obstruct the views.

Confusion of values — Some of the confusion of natural and artificial values in the 1887 plan that would lead to future policy problems could also be seen in the suggestions for treatment of the eastern meadow on Goat Island which was to be treated in a fashion similar to that of the mainland just discussed (see section on the eastern meadow). Prior to 1885, sections of the east area had been allowed to regenerate naturally after their disruption earlier in the nineteenth century. The plant community at this end was obviously not considered natural. Natural, apparently, was the interior woodland and the undisturbed islands rather than the equally natural successional plant communities, such as the meadow area in 1885. In the cultivated ground, and the "few small clearings" made here and there, possibly to accommodate certain outbuildings seen on early maps of the area, "a thick young growth has sprung up." Rather than permitting this natural regeneration, probably not understanding its relationship to restoration of the primitive environment on Goat Island, Olmsted and Vaux suggest that this area be thinned and planted such that "all these spaces shall be refurnished with trees in the manner proposed for the mainland but less closely, as the foliage is not intended as a screen, and some variety in its disposition, as shown on the drawing, will be pleasing."

Unfortunately, this treatment was little less artificial than the original total disturbance of the regenerating habitat by agriculture. What followed in this area was the burning of it clear, the ploughing what topsoil there was, depositing topsoil brought from off the island, and dynamiting holes in the bedrock for the planting of trees, all of efforts which took place by 1915. This radical treatment seemed to agree with one interpretation of the 1887 plan, with embellishments such that now a lawn with shade trees, bearing no relationship with the primitive landscape, is maintained here.

Olmsted and Vaux, although interested in their primitive landscape, apparently did not lend their thoughts to the natural processes by which such landscapes evolve and maintain themselves. They were seemingly so preoccupied with the idea of "restoration" that "regeneration," a process that, if properly controlled, would guarantee something not artificial, did not seem to occur to them. The vegetation itself, out of the intrinsic forces within it, will strive to perpetuate, propagate and expand itself. Such processes were somewhat recognized in the plan, such as the happy location of a native tree in a useful place on the mainland, or the Cave of the Wind shelter "partially covered by the trees and bushes that have sprung up just outside of the position" which would totally conceal the structure after a few years growth. Olmsted and Vaux urged the speedy building of the road "to take advantage of the healing processes of natural restoration by fresh growth, which will commence as soon as the opening required for the new roads are cut through the existing woods."

to Goat Island

view
upriver

to falls

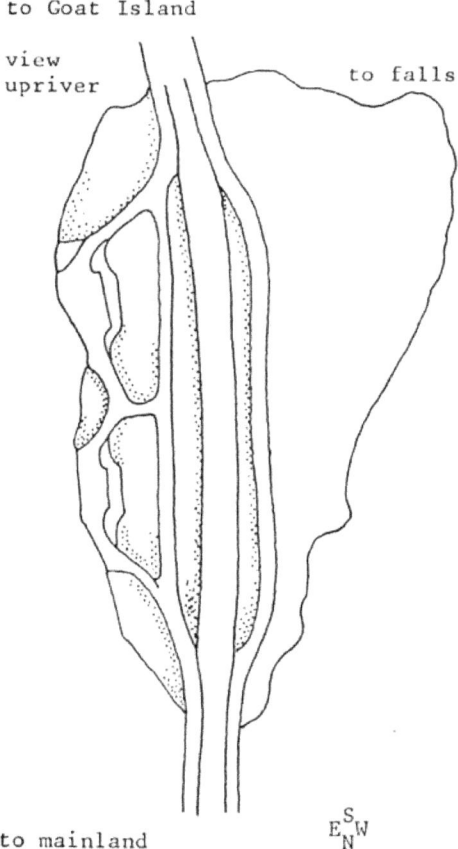

to mainland

Bath Island Plan — The diagram above is a "study for drive and vista-points" on Bath (Green) Island, ca. 1886, from the Manuscript Division, Library of Congress (Beveridge, 1985). There was apparently no attempt to reforest this island in this century—the aspect of 1917 is basically that existing today. Native trees naturally reseed themselves on the boundary, exotic trees have been established in the middle. The lovely view upriver is blocked today by the east (vehicular) bridge. Apparently the lawn exists as a monument to A. H. Green as a pun on his name (see text) when the original name (Bath Island) was changed to Green Island in his honor.

Green on Green Island — A. H. Green photographed on Green Island, November 14, 1898.

Schoellkopf on Green Island — At the same place Green was photographed above, a bust (removed after many years) was later placed honoring Paul Arthur Schoellkopf. The bronze plaque is still there. P. A. Schoellkopf was vice-president of Hydraulic Power Co. of Niagara Falls 1913–1918, and became president of Niagara Falls Power Co. in 1919 and president of Buffalo, Niagara and Eastern Power Corp. in 1925.

Tesla on Goat Island — A nine-foot bronze statue of Nicola Tesla, a giant in early electrical engineering, has been sited on Goat Island since 1976. It is a gift of the government of Serbia, but its placement is at a major viewing area in the Reservation. Behind is the Power Portal, the original arch entrance of the former Edward Dean Adams plant of the Niagara Falls Power Company. Electric power was first transmitted for a distance from this plant to Buffalo in 1896. A large statue of Tesla may also be found on the Canadian side in the Queen Victoria Niagara Falls Park, unveiled in 2006.

Land artificially added to Bath Island (later Green Island) by the Porters in order to facilitate expansion of their industrial and commercial structures on that island, was, according to Olmsted and Vaux, to be allowed to wash away and return to the natural boundaries. This was never allowed to happen, probably because of the importance of this island in the later securement of structures conveying utilities, services, conveyances and visitors onto Goat Island. The fact that this island was never re-vegetated is a curious monument to the Reservation's staunch defender, Mr. Andrew Haswell Green, for whom the name of the island was changed from Bath Island in 1898 to Green Island (16 Ann Rep Comm, 1900). "As the island is a sloping green lawn, the name of Green Island is doubly appropriate." The mandate of the original legislation to protect and restore the Reservation's native flora was overlooked on this conspicuous island as an ironic honor to the second President of the Commissioners of the Niagara Reservation, because his last name was "Green" (see discussion above).

As to the primitive woods itself—the most important of the plant communities due to their historic value (see 1879 report)—Olmsted and Vaux only indicate improvements which would gradually result from continuous proper care of the forest, the particular methods of which involve too much of expert detail to be profitably considered in this report." Although nothing is known to this writer about what "care" was meant, one wonders how this reference conforms to Olmsted's earlier characterization of the wild, untamed and hence, uncared for character of the primitive woods to be restored.

Restoration becomes secondary — Olmsted and Vaux, true to their success at constructing urban parks, appear to have become carried away by their interest in vistas—especially in an area supposed to be so circumscribed, where all the delights of nature are concentrated in such a tiny area. Restoration, especially as far as the mainland is concerned, seems to take a secondary interest to the mechanics of "staging" nature for the stroller and the rider. The views on the island and river verges are controlled by natural and artificial disturbance—"openings" in vegetation competing to use up all available light, naturally attempting to close. The hunger to "see" would triumph over vegetative regrowth.

Just as in Central Park, "delightful views," whether of the mainland or of Goat Island were promoted on many occasions in the ensuing century—either to deliberately create them, or as a palliative when policy, such as later when digging out elevations reducing visibility for automobile traffic on the island called for the removal of vegetation. The consequently exposed sky, forest and river was always considered to be "delightful."

Much of the objectionable destruction of the scenery of Niagara Falls on the Goat Island complex over the past century was not done to promote the safety of the increasing numbers of visitors to the Falls, but in the anticipation of their convenience. The other side of degradation was to impose schemes of beautification on the rare, unique, wild structure of the landscape, which was never described in detail or given the protection of a plan to preserve its biology.

As though conscious of the lack of specifics in the 1887 plan, "the Commissioners consulted at various times four distinguished landscape architects, Mr. Frederick Law Olmsted, Mr. Calvert Vaux, Mr. Samuel Parsons, Jr. and Mr. William S. Edgerton" (19 Ann Rep Comm, 1903). The year after Olmsted and Vaux devised their plan for the improvement of the Reservation, the Commissioners sought the technical advice of David F. Day, a local student of the flora, who prepared a catalogue of the species of vegetation growing not only on the Reservation, but throughout the park areas and areas that should have been park: the gorge of the Niagara River—knowing that the ecological relationships between these areas were close. "Old pictures, the recollections of old residents, and careful observations of the natural environment were employed to assist in restoring the landscape as nearly as practicable to its original aspect," "Denuded places were planted with trees, care being taken to use only indigenous varieties" (19 Ann Rep Comm 1903).

Above all, a plan for restoration and preservation requires a continuous commitment and study at the site to understand the vegetational dynamics and their relationship to the elements of the Olmsted and Vaux plan, such that the Reservation scenery will become again "the ripened fruit of long preceding forethought and of patient persistence in carrying out the organic purpose of a plan. It is looked upon ... as a wild fruit."

The architectural treatments on the island were in celebration of the native woodlands—they were meant to be as imperceptible as possible by duplicating forms and textures already present in the environment, the stone a substrate on which the living things in the woods could establish themselves.

Some idea of architectural taste in buildings erected on Goat Island may be imagined by the description of the Shelter Building, built in 1896 by the bridge to Bath Island, which was "an inconspicuous structure; and the graceful lines and warm color of the roof, the red-brown tone of the stonework, harmonize with the natural surroundings ... The walls are of Medina stone, with arched openings; the posts, of Georgia pine, have curved brackets, supporting the overhanging roof ..." (13 Ann Rep Comm, 1897, see accompanying picture in this report).

Ultimate value of the Plan — The plan of 1887 has been of profound value in keeping close control over preservation of the island's biological value by inhibiting development, even if it could not stop that altogether. It is, however, with respect to the biological objectives of preserving and restoring the scenery, only half a plan. A supplemental plan attempting to fulfill the environmental objectives of the movement to preserve the scenery at Niagara must be devised, and a careful restoration begun taking full advantage of the considerable knowledge that has been gained in the field of biology since 1887. It is within a context of a restored and regenerated complex of ecosystems for which Goat Island had received world recognition by natural historians during the nineteenth century that Olmsted and Vaux's design would become perfected, and a century of conflict between biological and developmental issues resolved.

It is through recognition and reestablishment of the natural processes of the central woods and other habitats in the complex that the aesthetic, spiritual and social values expounded by Olmsted and Vaux, insofar as they relate to the Niagara Reservation, can be realized.

The Map

There is no evidence that I could detect to indicate that Olmsted's views in 1886 on the objective of establishing the Reservation had changed from those of the report of 1880. The illustration accompanying that report indicating the aspect of the restored environment is still the visual objective of restoring the scenery, although several details might inherently prevent such a restoration from happening. The map accompanying the Olmsted and Vaux report shows: (1) the layout of the roads, paths, buildings and other structures and (2) the layout of the vegetation (restored).

Olmsted and Vaux had worked out a system of roads and paths on a gridded sketch (see illustration). This sketch was handed to a draftsman, P. R. Jones, whose name appears on the map as "delineavit," or "Jones drew it." It is a schematic kind of diagram in which the structural details were exact to specifications.

The vegetation was presented differently from that in the 1880 Gardner Report for technical reasons, among others. The vegetation is drawn much as it was drawn for the rendering of park plans. But this was not a park, it was a botanical preserve—consequently, it must be understood that the clear white areas between tree crowns *is not lawn-space*. In the case of the central forest, the canopy is actually closed, yet the map draws the tree crowns generally as not touching. One assumes that the drawing of these areas would be confusingly crowded if both herbaceous, shrub and tree layers were depicted. The east meadow area, on Goat Island, was not a lawn with shade trees and picnic tables. On the 1887 map, these white areas actually depicted the successional herbaceous community that existed there, and which was problematical to Olmsted and Vaux. The herbaceous layer was left white. Examination of the manner in which the primitive plant communities on Goat Island, which in fact existed at the same time as the map was drawn, are depicted indicates they are drawn in the same way the restored mainland was depicted: the map was a schematic rendering only. Both Goat Island and the mainland were to look together like the illustration of 1880 when restoration had been completed.

In the schematic representation of the 1887 map, there are tree canopies depicted with little central points representing main stems. All the trees appear to be deciduous and uniformly distributed in the primitive areas. Note that the canopies depicted are open. Also depicted are stipple marks which are meant to represent shrubs, or minor trees with small canopies, such as the "Dwarf Willows" or native willows of the system. At major carriage-viewing areas on the mainland, such as at the Old French Landing, prospect areas are maintained with only shrub communities, as suggested in the body of the text. The end of Fourth Street in the village where the street entered the Reservation, perpendicular to the park's east-west axis, was also treated to a shrub-community, the better to offer a prospect through the framing tree canopy. These shrub communities are interspersed all along the river-margins, some occurring naturally according to the natural ecological characteristics of the river margins, some deliberately maintained.

Note the buildings in Prospect Park, in the Upper and Lower Groves, corresponding to the natural terraces there. These were the public shelters and maintenance and administrative facilities which were not to become established on Goat Island.

Note the arrows on the maps showing vistas, such as the disembarking area at the junction of Falls and Canal Streets before a building called the Acception House where a view up the American Rapids was to be prepared. All down the line of the arrow, the vegetation is drawn, according to the scheme, as shrub layer, as were the laybys along the paths just upriver from the Goat Island bridge.

The Acception House was presumably the structure where visitors were to picnic or otherwise refresh themselves before going out to the islands. This structure was to face the commercial enterprises of the village directly across Canal Street. Note that the city approaches to the Reservation at Prospect Park were planted to trees to prepare the visitor for leaving the urban experience behind. These appear to be the typical "shade-tree" alignment typical of most urban streets, with trees planted singly, at even intervals, in a row. The transition from city to park took place *in the village and not in the Reservation*. Ingeniously, the excursionists, who, en masse, constituted the "throngs," cooped up in trains for hours, disembarked onto Falls Street and walked, exercising away their tensions up a tree-lined avenue to the Acception House, pausing in the circular area junction with Canal Street to receive a tantalizing look up the American channel—not down toward the American Falls and Prospect Point. From there the visitor could walk directly to the Goat Island bridge, or take a Reservation carriage, or break out the picnic basket, buy a Reservation map and plan the day's events.

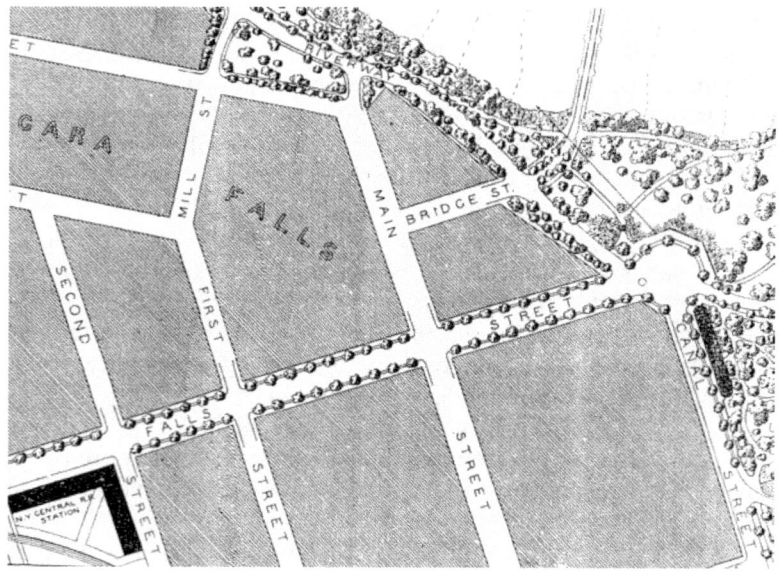

Plan map, excursion — Visitors buying excursion tickets to Niagara Falls and confined to trains for hours would disembark at the New York Central Railroad Station on Falls Street (lower left). As they approached the Reservation, they were directed toward Prospect Point by the rows of street trees regularly planted the three blocks to the grove areas and the Acception House, read information posted there and at the Railroad Station. Maintenance and administration buildings were located here. The Reservation carriage service and other services could be availed of here. The transition from city to park took place in the city, and not on park grounds. Visitors became oriented by "arboreal guides" beginning with trees established directly across from the exit from the railroad station. (Map from 3 Ann Rep Comm, 1887.)

Plan map, eastern boundary — The eastern boundary (upriver) of the Reservation ended at Port Day, the inlet canal for the hydraulic canal. A shelter was intended here, and a broad view of the upper river, from an area of historic significance. Note the character of the plantings depicted, showing shrubby river margins, and shrub or other low-growing vegetation in the prospect, or vista areas. The pedestrian path was flush with the carriage road, with diversions or loop paths and shaded seats. (3 Ann Rep Comm, 1887.)

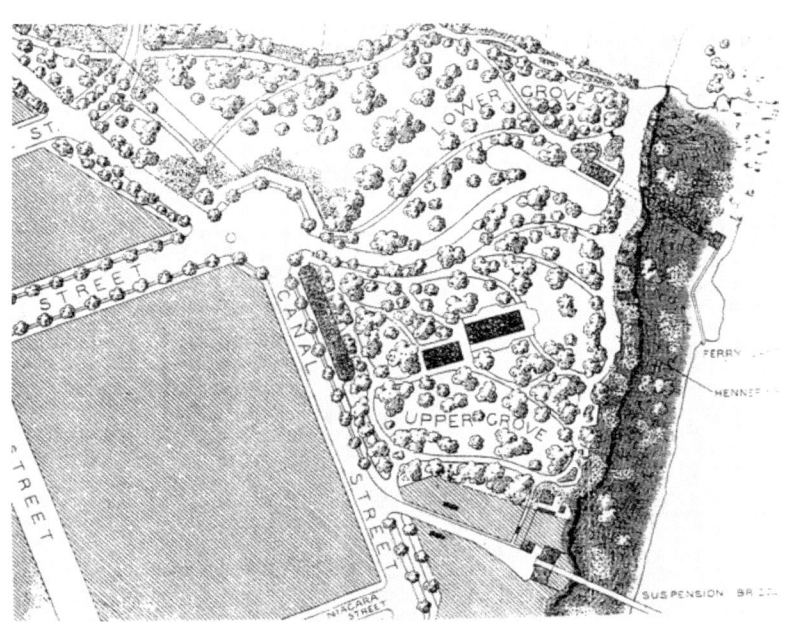

Plan map, Prospect Point — Revegetation of the Prospect Point area (upper and lower groves) was problematical to Olmsted and Vaux due to the poor condition of the trees there in 1886. Note the extensive areas opened on the gorge rim in the prospect areas. Most of the visitors to the Reservation were expected to be concentrated here, as were most of the functions of the administration. Large numbers of people were expected to congregate at the inclined railway down to the Ferry Landing below the American Falls. Trips to the Canadian Park could be had here, especially for those who came on the train. (3 Ann Rep Comm, 1887.)

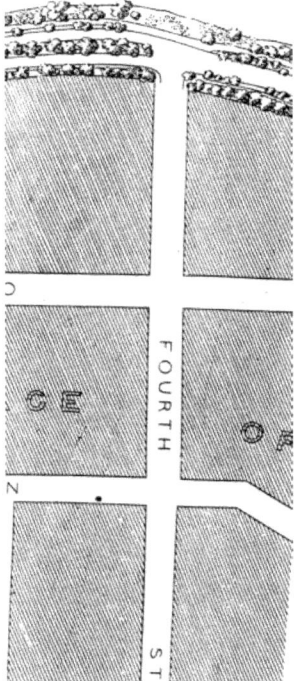

Plan map, vista — Vehicles traveling down Fourth Street entered the Reservation on the Riverway where a vista of the American Rapids was provided by the installation of low vegetation, framed by forest trees. (3 Ann Rep Comm, 1887.)

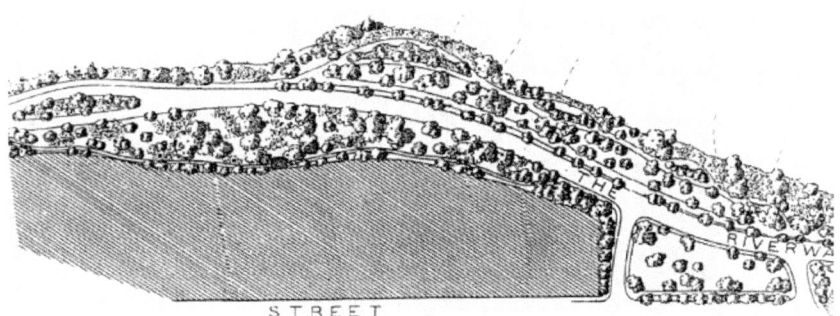

Plan map, Riverway — The complexity of street and path systems for the mainland is shown here. A small path runs along the top of the terrace at the back boundary of the Reservation where a good view of the river could be had. Loop walks blocked from one another by vegetation were numerous. (3 Ann Rep Comm, 1887.)

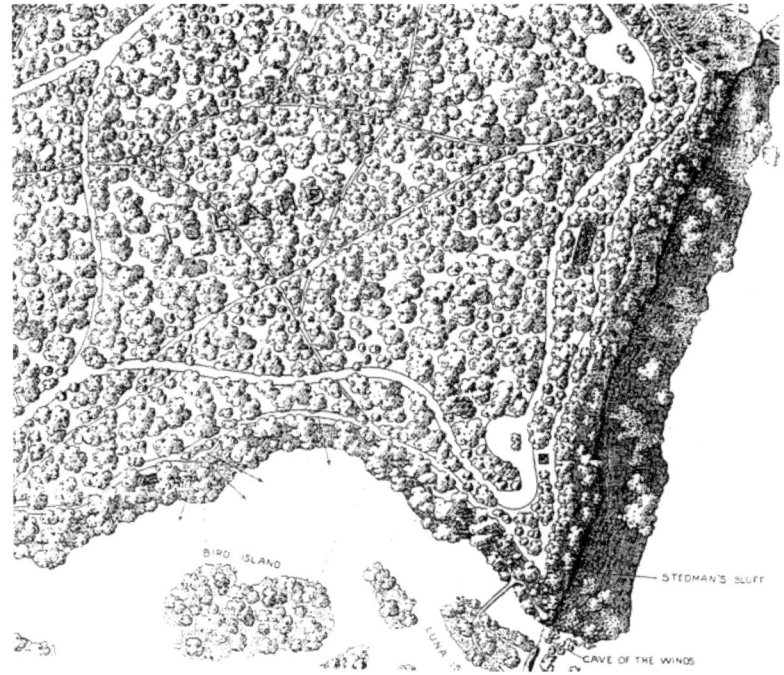

Plan map, Goat Island at American Falls — This area was part of the most important section of the Niagara Reservation, for it supported the primitive environment of management of other sections of the Reservation was supposed to copy. The whole Reservation should have matched the aspect of this area through time. As for the interpretation of the map accompanying the third annual report of the Commissioners of the Niagara Reservation (1887), the character of this area is to be compared with the character of areas to be restored elsewhere in the Reservation, such as on the mainland. White areas do not depict lawns, but only the lowermost layers within the structure of the primitive plant communities: the layer of herbs (as opposed to shrubs and trees), of mosses and liverworts, fungi, algae not associated with trees and shrubs. There is no indication on the Olmsted and Vaux maps that lawns were to be established anywhere in the Reservation. The arrows represent prospects for viewing. (3 Ann Rep Comm, 1887.)

Plan map — The eastern meadow of Goat Island was permitted to maintain its open characteristics. Note one of the two shelters permitted on the Island (dark ellipse), according to the Plan. A large parkage, or lay-by, was placed at the extreme east end for a broad view of the upper River. (3 Ann Rep Comm, 1887.)

ANDREW HASWELL GREEN AND OLMSTED

Although the Niagara Reservation was one of Frederick Law Olmsted's chief contributions to American culture, he was not to enjoy the fruits of this triumph, nor, as a father his growing child, see to its development and maturity.

The conflict that was to exclude Olmsted from his Niagara is usually ascribed to events in the history of Central Park and the antagonisms in New York City between public and private interests. New York State and New York City had their political problems, and one man who facilitated progress in the arena of great civic monuments in spite of these difficulties was Andrew Haswell Green of New York City, Comptroller of Central Park, protector of the interests of the municipalities comprising the metropolis of New York City. Governor Cleveland, perhaps recognizing Mr. Green's ability to defend the public interest and the personal altruism and sacrifice Green was willing to demonstrate for the public good (see Eulogy, 20 Ann Rep Comm, 1903) gave him an appointment on the Commission of the State Reservation, which he held for twenty years, fifteen of which as its President.

Mr. Andrew Haswell Green (see page 93) had figured prominently in the development of Olmsted's career. When the design for Central Park by Olmsted and Vaux was accepted and Olmsted's career began its meteoric rise, Green was appointed controller of Central Park. It was through Mr. Green's administrative ability that the ideals of Central Park were brought forth into reality in the political atmosphere of what was to become the incorporated City of New York.

"[Central Park] gained its strongest administrative defender in 1857 when Andrew Haswell Green was elected Treasurer of the Park Board. A lawyer by profession, closely associated with the Democratic leader Samuel J. Tilden, Green was one of the outstanding New Yorkers of the last century. Endowed with unusual energy and a Roman sense of civic responsibility, he let little escape his attention in matters beneficial to the city. The designers of Central Park were to have many differences with Green, but it was he who fought the battles in the noisy political arena," (Reed & Duckworth, 1972).

Certain aspects of the conflict between Frederick Law Olmsted and Green had to do with conflicting styles of management. "Olmsted's concern for [Central Park] went far beyond concern with Park-Keepers; he wanted control over every detail of management as well as design.

The control he demanded led, of course, to conflict with the Park Commissioners, especially with the park's financial watchdog, Andrew Haswell Green" (Reed & Duckworth, 1972). "In 1863 both he and Vaux handed in the first of several resignations, so strong were their differences with the Park Commissioners. A year later Olmsted still chafed at the memory of what he called Green's 'systematic small tyranny ... he was slow murder.' Yet in 1865 Green was largely responsible for the designers' being rehired as landscape architects." (Reed & Duckworth, 1972). Whatever the nature of Green's tyranny, Olmsted referred to it, in 1863, as "Greenism" (Roper, 1973).

Green, when comptroller of Central Park, exercised absolute control over park expenditures. However, "[he] delegated none of it, put no confidence in his subordinates, and dominated in most instances the judgment of his co-commissioners through his superior knowledge of park affairs. ... he was rigidly honest. He was also rigidly stingy ... if he had brought to bear on a family his strong constitutional distrust of all honesty and capacity but his own, he would have made life as miserable for his dependents as he did for his park subordinates" (Roper, 1973). Green ..." cracks his whip over good & bad indiscriminately, and is not generally beloved therefore ... (George Waring, in Roper, 1973). Green, too, however, had his hands full with Olmsted, whose personality he referred to as "overwhelming." Central Park, existing due to and at the mercy of New York City government, Calvert Vaux wished to see as "the big art work of the Republic," an "artistic success, not an organizational and executive success" (Roper, 1973). "I have always felt that it would be mean on the part of its makers to let the success be an administration success. It would seem as if they were ashamed of their work" (Vaux, in Roper, 1973). Yet Olmsted's managerial talents were not his strong point, as acknowledged by his partner. His "administration had not been comprehensive, calm, or statesmanlike, and his diplomacy had been 'very defective and impatient" (Vaux, in Roper, 1973).

"Boss Tweed succeeded in driving Andrew Haswell Green out of office in the spring of 1870, and the two landscape men were forced to resign that fall .... With Tweed's downfall in 1871 the two designers were back again, horrified at the destruction done [to Central Park] in the name of correcting 'neglect.'"(Roper, 1973).

Details of Mr. Green's life are given in the eulogy (20 Ann Rep Comm, 1904) at his death by murder on November 13, 1903 "by a crazy man" (Green, 1905). He was a "member of the board of education, as the controller of the Central Park, of which more than any other one man he was the creator; as the controller of the city's finances at a time when immovable integrity was required to stay the riot of municipal plunder; as the projector of great institutions of art and learning, and of parks and other material improvements for the health, comfort and pleasure of the people" (20 Ann Rep Comm, 1904). At 37 years of age, Green became a Commissioner of Central Park, "for which the movement had begun some years before." "In 1857 he was appointed by the Legislature one of the Board of Commissioners of the Central Park and became treasurer of the board: president and executive officer of the board, that is, controller, of the park for about ten years. He had the complete supervision of the engineers, landscape architects gardeners and the whole retinue of employees, sometimes comprising as many as 3000 men." Mr. Green "by his care for Central Park, was led to care for related enterprises, as the Museum of Art, the Museum of Science and the Zoological Garden," (20 Ann Rep Comm, 1904).

It was Andrew Green, as Deputy Comptroller of New York City, who provided the financial figures that, used as legal evidence, proved corruption in New York City government, breaking the "Tweed Ring" then extorting millions from city government, and established the basis for reform, although "the immediate cause of Tweed's downfall was the publication in the New York Times of evidence of wholesale graft revealed by M. J. O'Rourke, a new county bookkeeper" (Bridgwater & Kurtz, 1963). As comptroller of Central Park, Green made the recommendation for the consolidation of many towns, villages and cities into "a great municipality entitled the 'City of New York'," which was later accepted and accomplished (20 Ann Rep Comm, 1904; for additional details of Mr. Green's accomplishments see Green, 1905).

Green's miserliness, so exasperating to his colleagues, was actually the source of his effectiveness as a politician and political reformer. Bookkeepers, unsung heroes of the American justice system, such as those working with the Federal Bureau of Internal Revenue, also brought to an end the crime syndicate of Al Capone in Chicago in the 1920's.

By 1873, "Green was one of the most powerful politicians in" New York City. "Shrewd and experienced, he had developed strong public support based on his impregnable reputation for honesty. With the collapse of the Ring, he had been appointed deputy comptroller, and then comptroller, of New York City. The city's finances were in Augean disorder, and with Herculean energy he set about policing them. Economy was his passion .... Both hated and admired for his conduct of his office, he seemed indifferent to his popularity. Dedicated to economy, he delighted in power; his fiscal authority was the weapon he brandished in the service of both" (Roper, 1973). Green's comptrollership ended in

1876. As far as New York City was concerned, Green was permanently out of office. "Two years later Olmsted was dismissed, and in 1883 Vaux resigned." These events lead to Olmsted's break with New York City, as discussed above.

Green, considered a "reform Democrat," that is, anti-Tammany Democrat in the Tammany-dominated City of New York (Roper, 1973) was familiar with dealing with large corporations and financial associations that sought to aggrandize themselves at the expense of municipalities. The upcoming conflict between preservation of the Niagara Reservation, involving the existence of Niagara Falls itself and private corporations in the village of Niagara Falls during Green's Presidency of the Commission of the Niagara Reservation appears to have been similar to his successes in New York City—even to preserving the waterways of that City. "In a striking way he treats of the water system, and shows that though commonly looked upon as designed by separate populations one from another, it really and effectively unified them. He portrays the spoliations and injuries that are committed against the navigable waters of New York and especially against the shores bordering upon them. Both the land and the water are wrongfully employed by marauders, from the giant corporations, who break the commandment 'Thou shalt not steal,' down to the obscure captain of a garbage boat or mud scow" (20 Ann Rep Comm, 1904). Green's sensitivity to the role control of the waterways was to have on the appropriate means of governing geographic areas would be well displayed at Niagara. He recognized the fundamental dependency the falls of Niagara had on the volume of water upstream, such that he initiated and saw enacted federal legislation which was to ultimately lead to establishment of what is now known as the International Joint Commission (15 Ann Rep Comm, 1899).

Green and the Reservation — Green, too, had been an advocate for the establishment of the Reservation at Niagara through association with his law partner, another powerful reform lawyer and politician: Samuel Tilden. Tilden instituted reforms in 1872 which were meant to check the "flagrant abuses under the reign of William M. Tweed" under the Tammany administration of New York City, and managed to have Tweed tried on felony charges (Bridgwater & Kurtz, 1963). Tilden "may ... have encouraged [Governor] Robinson" to direct the Commissioners of the State Survey toward preparation of the Gardner report" (Kowsky, 1985). Tilden again was said to have had influence over Governor Hill, who signed the 1885 bill into law, as Tilden was "the political mentor of

David B. Hill, who was Tilden's foremost pupil and disciple" (Welch, 1903).

Green's advocacy of the Niagara Reservation was not, however, exerted through the Niagara Falls Association, a society in which Olmsted and his friends predominated. Green's name did not appear in the roster of members (Executive Report of the Niagara Falls Association, 1885).

Green was founder and President of the American Scenic and Historic Preservation Society, a Commission of the State of New York established in 1895—ten years after the Niagara Reservation was established (Newton, 1971). This Commission was to promote the development of New York State's park system through the acquisition of Watkins Glen, the founding of the Bronx River Parkway Commission, Letchworth Park, and the ultimate votes in 1910 and 1916 by the people of the State "to approve a bond issue of $2,500,000" and $10,000,000 respectively for the purchase of land for a state parks system. The successor in State government was to become the State Council of Parks under the Department of Conservation, chaired by Robert Moses (discussed below).

When the Commissioners wrote in 1891 of exciting gains in the public land movement of the time, there also appeared what may be interpreted as a pointed reference to Yosemite during the tempestuous period near the year of its return to federal control: "It has been recently asserted that the management of the wonderful valley has not been as careful, intelligent and vigilant as it should have been, and that the scenery has consequently suffered injury" (7 Ann Rep Comm, 1891). Olmsted had been, after all, leader of the commission appointed by the California governor to protect that valley, much as Green himself was later head of the commission to protect the natural scenery of Niagara. Even a recent treatment favorable to Olmsted's extraordinary contribution to American culture could not suppress a hint of apology for Olmsted's essential abandonment of Yosemite, for "at any rate, under the circumstances he could not be blamed for California's half-hearted management of the valley in ensuing years" (Newton, 1971).

Green's actual objections to Olmsted's concept for the Niagara Reservation, either in the Gardner report of 1879 or the later report Olmsted co-authored with Vaux, have not been detailed in the literature I have seen.

In reviews of Green's life written at the time of his death, Green's association with Olmsted is little if anywhere mentioned, and his exertions on the administration of Central Park are emphasized such that

Green himself appears to have been the creator of that Park, the first major urban park, which was to become an American institution. The Niagara Reservation may have lost much subsequent prestige because of this conflict, and the magnitude of its cultural contribution thus lessened in the continuing history of public land policy in the United States.

It was Green and state politics that were to dominate Reservation affairs in the decades ahead—with much that was to prove potent to the protection of the young preserve then and into this century. While Olmsted may be seen as someone interested in the "social, human, and aesthetic problems" associated with urban growth, Green "was almost exclusively concerned with its physical and administrative development" (Roper, 1973).

Although unwilling to work with Olmsted himself, Green was anxious to work with his associates, indicating that, although there may have been a conflict in personality, they had much in common on a conceptual level.

"As for realization of the plan [of 1887], Olmsted was willing to leave the construction work to Vaux and his partner George Kent Radford. Vaux was also eager to provide a professional opportunity for his son Downing Vaux, an architect" (Beveridge, 1985). Calvert Vaux designed the "simple, incurving three-railing guard rails," the bridge to Luna Island (1894) "and apparently made the plan for the bridge to the First Sister Island in the same year. After Vaux's death, his son Downing continued to provide plans for buildings and bridges [on the Niagara Reservation], while George Radford did further engineering work and another of Vaux's partners, Samuel Parsons, Jr., continued to supervise landscaping and planting" (Beveridge, 1985). Note that the Vaux bridge to Luna Island was replaced by the "modern concrete structure" visible today, in 1933 (23 Ann Rep for 1933, State of New York Conservation Department, 1934).

Throughout the early annual reports of the Commissioners, references were made to contracts only with Vaux and his associates, mainly construction-related: $45.00 to Downing Vaux for a "tracing map," (1 Ann Rep Comm, 1885), $50.00 to Downing Vaux for a map or mapping (6 Ann Rep Comm, 1890); $309.20 to Vaux & Co., for surveying (9 Ann Rep Comm, 1893); plans were submitted for a bridge to Luna Island which had been wood (10th Ann Rep Comm, 1894), and of a stone bridge to the First Sister. "Iron guard railings of the design proposed by Mr. Calvert Vaux have been erected on the retaining walls" perhaps "at the entrance to the bridge to the islands," (10 Ann Rep Comm, 1894); $410.81 was paid to Vaux & Co., Luna Island bridge plans, and $72.75

for the traveling expenses of Downing Vaux, (11th Ann Rep Comm, 1895); $125.00 was paid to Vaux & Emery for plans and specifications, another $300.00 later in the year for the same return, (12th Ann Rep Comm (1896). $176.36 was paid to Vaux & Emery, shelter building, and $74.25 to Marshall L. Emery and $51.75 to Downing Vaux for traveling expenses, $146.48 was paid to Vaux & Emery for "terminal station" (13th Ann Rep Comm, 1897). $147.50 was paid to Vaux & Emory, Hennepin's View and buildings, later $75.00 for First Sister Island bridge, (16th Ann Rep Comm (1900).

"Six hundred and twenty-five feet of iron railing of the pattern designed by the late Calvert Vaux, have been erected between the Cave of the Winds and the Horseshoe Falls. The standards are set on posts of Medina sandstone, four feet long, embedded in the ground. The iron railing now extends from the American to the Horseshoe Falls," apparently along the crest of the gorge on Goat Island (13 Ann Rep Comm, 1897). Vaux's iron railing was established on Luna Island in 1900 (17 Ann Rep Comm, 1901).

Professional consultation to Superintendent Thomas Welch regarding the nature of the vegetation to be established at Prospect Point was made by Theodore Wirth, a landscape gardener, "after consultation with Samuel Parsons, Jr., and were planted by Mr. Wirth under the direction of Mr. Parsons" (12 Ann Rep Comm, 1896). Samuel Parsons Jr. was a partner of Calvert Vaux and came from a family of nurserymen. He became landscape architect to the City of New York sometime before 1898, and before that had served as superintendent of planting in Central Park (Newton, 1971).

Frederick Law Olmsted, Sr. retired by 1901 (Newton, 1971), but his son, Olmsted, Jr., was to continue in his preservationist tradition at Niagara Falls.

"Property holdings of the Niagara Falls Power Company and its filial companies before the consolidation of 1918" — On a fold-out map of power company land in Adams (1927), one can see the areal relationships of parks and dedicated hydroelectric power facilities. Portions of this map are presented here, from right to left on the map.

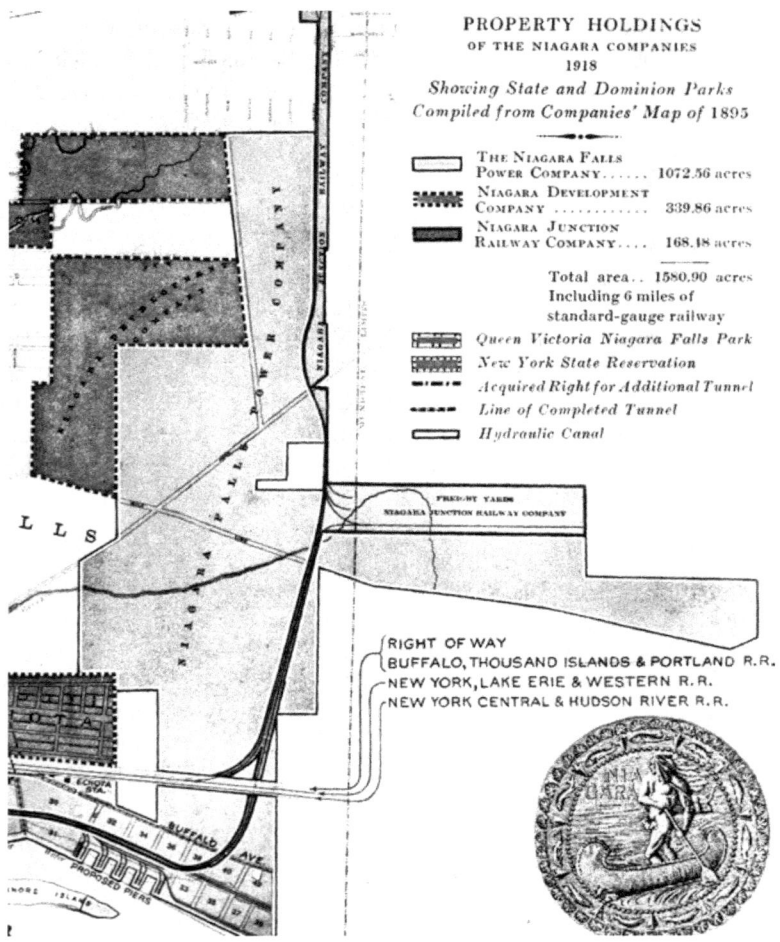

PROPERTY HOLDINGS
OF THE NIAGARA COMPANIES
1918
Showing State and Dominion Parks
Compiled from Companies' Map of 1895

| | THE NIAGARA FALLS POWER COMPANY...... | 1072.56 acres |

NIAGARA DEVELOPMENT COMPANY .... 339.86 acres

NIAGARA JUNCTION RAILWAY COMPANY.... 168.48 acres

Total area.. 1580.90 acres
Including 6 miles of
standard-gauge railway

Queen Victoria Niagara Falls Park

New York State Reservation

Acquired Right for Additional Tunnel

Line of Completed Tunnel

Hydraulic Canal

RIGHT OF WAY
BUFFALO, THOUSAND ISLANDS & PORTLAND R.R.
NEW YORK, LAKE ERIE & WESTERN R.R.
NEW YORK CENTRAL & HUDSON RIVER R.R.

1. Power company land — The area owned by the Niagara Falls Power Company and the affiliated Niagara Development Company and Niagara Junction Railway Company away from the Niagara River was extensive.

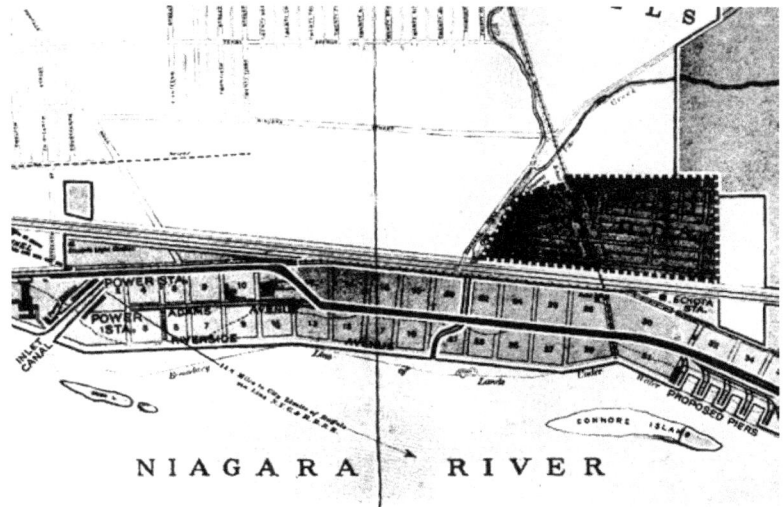

2. Power company land along river — Most of the remaining land along the Niagara River upstream of the Reservation was owned by the Niagara Falls Power Company.

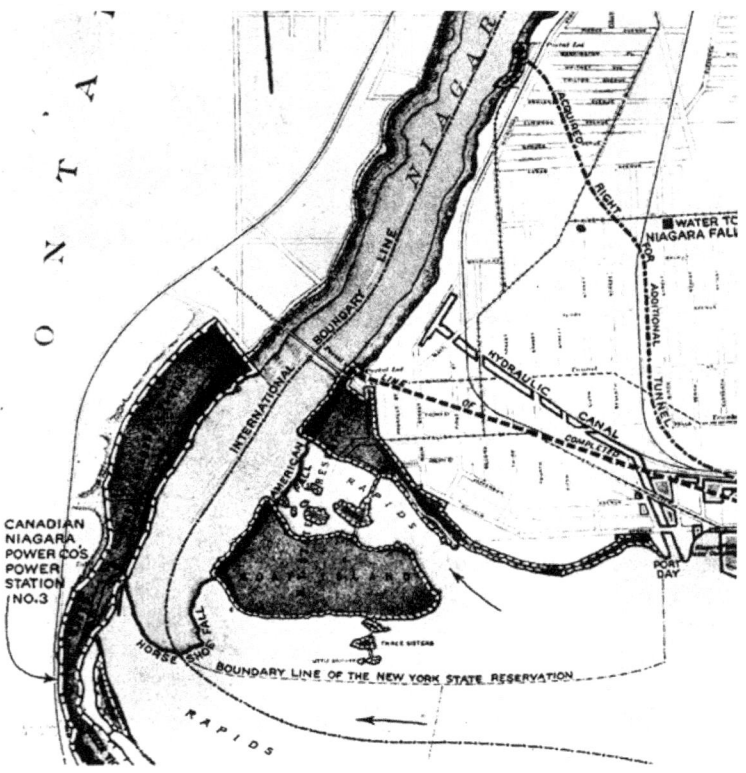

3. Niagara and New York park areas — On the left is the Canadian Niagara power station situated in the Queen Victoria Niagara Falls Park (white dashes). The Reservation on the New York side of the border (white dots) consists of Goat Island, Prospect Point, the islands between, and a length of shoreline reaching to the Hydraulic Canal inlet, used for mechanical power. The straight black dotted line is the tunnel leading to an early power station on Reservation land. The curving black dotted line is an acquired right-of-way for a tunnel to what will become the Schoellkopf power station.

THE COMMISSIONERS OF THE STATE RESERVATION AT NEW YORK:

ANNUAL REPORTS TO THE NEW YORK STATE LEGISLATURE

The Green Administration

The year after publication of the Olmsted and Vaux plan for improving the Reservation (1887), under Dorsheimer's presidency, it was reported that the request for funding for such improvements was denied by the Legislature after review in the Finance committee of the Senate and the Ways and Means committee of the Assembly (4 Ann Rep Comm, 1880).

After the financial review, the Commissioners insisted that expenditures be made on the Reservation according to an accepted plan that would be backed by the financial means to implement the plan, rather than allocate money in a piecemeal fashion according as necessity arose, such as spot repairs on existing facilities. Erosion on Goat Island was serious, and the roads and paths were disintegrating on the south shore. Bridges and other structures were considered unsafe and needed replacement.

In Ontario, the establishment of the corresponding Canadian park was taking place. In April of 1887 the "House of Assembly of the Province [of Ontario] passed an act authorizing the issue of bonds to the amount of $525,000 to purchase and improve lands to be known as "the Queen Victoria Niagara Falls Park." The bonds, bearing four percent interest and guaranteed by the Provincial government, were immediately purchased. Of the proceeds, $402,867 was required to meet the awards of the arbitrators of land appraisal, while more than $100,000 remained available for work of improvement. "Of this sum about $26,000 has been expended, with excellent judgment, during the past season [1887], the result being that the lead in the noble project of an International Reservation at Niagara may now be said to be taken by our Canadian neighbors" (4 Ann Rep Comm, 1880).

Again, the environmental plea is repeated for restoring the shoreline by the American Falls "by planting this strip of land with trees" ... "the whole village [of Niagara Falls] may be shut out from view—planted out—and the unsightly walls, the sewer mouths and wing dams, replaced by natural banks like those of Goat Island." The attribution of these remarks was to James T. Gardner, referred to as "Professor." There was to be a healing process "of time and nature."

The Commissioner's "first care" "will be to enlist all the friendly energies of nature to aid in drawing around the great cataract some semblance, such as art may create, of that magic circle of sylvan wildness and seclusion within which Hennepin discovered it," that is, to restore the environment by an interaction of natural forces and human design. It is apparent in the fourth annual report that the money they desired of the Legislature was twofold: 1. that the taxpayers, "the people who pay the price shall be permitted to enter into the full enjoyment of the purchase" with improved roads, paths and structures, and 2. to "leave to posterity to gather the rich aesthetic fruit of their planting," that is, the restoration.

But the main issue of 1887 was disappointment in the failure of the Legislature to appropriate adequate money to carry through the reasonable plan for the permanent administration of the Reservation.

In an attempt to forestall the development of a railroad, disfiguring the lower banks of the Niagara River gorge, the Commissioners wrote that "it would be most desirable, and it may still be practicable, for the State to secure the line of the slope in question from its present boundary [at the north and west end of the Reservation, that is, at the head of the gorge] to and including the whirlpool." The effect of this extension of the Reservation boundaries "need involve no interference with the important industrial and manufacturing developments on the river bank below the American Falls; while it would render the reservation of Niagara's essential surroundings complete." They cite the action on the part of the Ontario and Dominion governments "by which the entire bank and debris slope on the opposite side as far as Queenston will be included in the Canadian park, [which] makes a corresponding acquisition on this side of the river doubly to be desired."

It is apparent that the Commissioners were not aware that in 1886, the year after the Reservation's establishment, the Niagara Falls Hydraulic Power and Manufacturing Company had "secured a deed for the slope or strip of land between the high bank and the lower river," as mentioned above, which was sold to the Company by the Porter family—an area "not included in the original grants acquired from the Porter family" (Adams, 1927). They must not have known of this sale of land because of the statement made above that the acquisition of the slope would not affect business on the upper slopes—which would have occurred if they had tried condemning this property by right of eminent domain. They may have been unaware of the technological advances being made which increased hydraulic efficiency that allowed full utilization of the whole slope, not just the upper fifty feet or so.

The Hon. David F. Day, of Buffalo, New York, provided a list of the flowering plants growing in the vicinity of Niagara Falls encompassing the flora of the entire Niagara gorge area on both sides of the river, as a tribute to the "rich flora of the region having especially attracted notice from scientists, both American and foreign."

The fourth annual report was signed again only by Dorsheimer, Anderson and Rogers.

5th Annual Report of the Commissioners of the State Reservation at Niagara for the fiscal year from October 1, 1887, to September 30, 1888. Transmitted to the New York State Legislature February 6, **1889**.

The next year, 1888, Green assumed the Presidency of the Niagara Commission, an appointment he would retain until his death in 1903. New names appear as Commissioners: James Mooney (Buffalo), John Hodge (Lockport), John M. Bowers (New York City), who were all appointed May 11, 1888 (19 Ann Rep Comm); Treasurer and Secretary is Henry E. Gregory (New York City, appointed Dec. 29, 1887), that is, none of the original Commissioners remained on the board except Green.

Green began his administration with strong convictions and a high moral tone: the grandeur of the spectacle of Niagara Falls in "its unobstructed contemplation was a high moral benefit to the community, and that the consciousness of its neglect and of its practical destruction, as a natural spectacle, would be morally injurious to the people" (5 Ann Rep Comm, 1889).

Green appeared to see the falls as a holy place desecrated by commercial interests. "Men influenced only by the spirit of trade and personal advantage had wrought an injury to the scenery of Niagara, not to repair which, would be a neglect of public duty. It is the privilege of the State hereafter to guard the lands of the Niagara Reservation from profanation." Green braced himself against his opponents: "the utility of the swift-rushing stream to owners of mills and factories was early recognized. Useful as mills and factories are, they are never beautiful, and the presence of a number of such architectural deformities near the Falls could not fail to strike the sense of all lovers of nature as incongruous. But the dominant materialism of the age, refusing to spare even Niagara, exacted service of the river, and extorted therefrom a tribute to commerce."

Request was made of the Legislature for adequate funds to prevent serious erosion then occurring on the south side of Goat Island, which, "with its varied wealth of trees and shrubs ... is the garden of the reservation." Money was also requested for paths, roads, grading, sodding, filling and planting.

During the year (1888), the Queen Victoria Niagara Falls Park was opened to the public. It was noted that that park was granted more extensive lands into its boundaries than the American one. It is at this point that the Commissioners urge the construction of a bridge to Canada "at or near the Whirlpool bluffs" downriver. Green contacted Colonel C. S. Gzowski, Chairman of the Commissioners of the Canadian Park in this regard, who was, on August 18, 1888, in England.

Gzowski was the first Chairman of the Canadian Commission, serving from 1885 to 1893 (Seibel, 1985). As a civil engineer, and Superintendent of the Public Works Department of Upper Canada (Seibel, 1985), Gzowski would know how to build a bridge. In a private capacity he had been "engineer for the building of the International Railway Bridge across the Niagara River between Fort Erie and Buffalo" (Seibel, 1985), that is, Black Rock, which was built in 1873 (Greenhill, 1985). This was a bridge built by the Grand Trunk Railway, which would become a major competitor with Vanderbilt's New York Central for the run to Chicago through southern Canada (Greenhill, 1985). Gzowski's private railroad connections included joining "a private contracting firm which built the Grand Trunk Railway between Toronto and Sarnia" (Seibel, 1985).

In this communication to Gzowski, Green urged extension of both the Canadian and the American park north to include the Whirlpool, below which the bridge was to be built, to facilitate the movement of people throughout all natural areas of interest, and thus unite the parks and fulfill the larger plan developed by Olmsted and Gardner in 1879.

Gzowski, in his reply, referred to the Canadian park as "the result of the efforts made by my colleagues and myself, towards the restoration of the scenery on the Canadian side of the Niagara Falls."

In a quote by Tyndall, the objectives of both governments were recognized, that "Niagara should have been considered so interesting and important as to induce the two governments to assume the ownership of lands circumjacent to the Falls, in order that the scenery, restored to its primeval beauty, might afford instruction and give delight to mankind."

The Commissioners of the American park write confidently that the two boards of Commissioners will be in "substantial agreement ... as to the course to be pursued in the improvement of both sides of the river.

In the work of restoration and development it is to be expected that both boards will recognize the importance of progressing along parallel lines, keeping constantly in view essentially the same ends, clearly perceiving the value of harmony and congruity in the final results." After saying this, the Commissioners appear to declare for the benefit of the Canadian board the objectives to which they were mutually to adhere, such as intolerance of "garish or glaring structures" and that the "simplicity of nature should at all times be preserved."

By 1889, the only buildings in the Reservation were those in Prospect Park, an office building on Bath Island, the cottage on Goat Island at the end of the bridge to Bath Island, the Cave of the Winds building and the pavilion associated with it. The roads on Goat Island are "common dirt roads", the paths made of gravel.

6th Annual Report of the Commissioners of the State Reservation at Niagara for the fiscal year from October 1, 1888, to September 30, 1889. Transmitted to the New York State Legislature January 22, **1890**.

In the sixth annual report of 1890, for 1889, the Commissioners are: James Mooney (Buffalo), John Hodge (Lockport), John M. Bowers (New York), Daniel Batchelor, with Andrew Haswell Green as president. Mr. Batchelor derived from Utica and "had an extensive acquaintance with agriculture, and was especially familiar with trees, shrubs, grasses and seeds, upon which subjects he was regarded as an authority. His suggestions concerning the plantations on the reservation could not fail to be valuable" (eulogy upon his death, 1893, 11 Ann Rep Comm, 1895).

Again, the Commissioners wanted to get on with restoring the scenery to its natural condition "and to preserve it from possible injury and violation as long as it should remain under their care and subject to their control." For two years the Legislature, although providing for care and maintenance, denied the Reservation enough revenue to effect "laying out the grounds and restoring the scenery of the Falls to its natural condition," according to Green.

Two kinds of visitors, one enjoying the paths, the other the roads are referred to in an appeal to improve both facilities. An appeal is made for an elevator to replace the Biddle Stairs in addition to a stairway. There is no money for turf and trees for the mainland shore. Present bridges are to be replaced by more permanent structures.

In March, 1889, "there was introduced into both Houses of the Legislature a bill entitled 'An act to authorize the Niagara Hydraulic Electric Company to erect machinery under Niagara Falls for the purpose of utilizing the water power of said Falls for manufacturing electricity and to erect the necessary machinery for the same."' This company directed its interests toward Prospect Park where they intended to divert the water falling over the brink there. A vault in the rock would be blasted away behind the curtain of falling water and in this cavity the company "proposed to place dynamos to be operated by water descending through a tube or well upon turbine wheels."

It is clear the attitude taken by the Commissioners. They "will decline to entertain propositions or applications, on the part of individuals or corporations, to utilize the water power at Niagara: nor will they countenance any scheme or project the success of which would be likely to result in defacement of the landscape, or in any way interfere with the performance of the duty entrusted to them, namely, that of restoring the scenery to its natural condition."

The Niagara Hydraulic Electric Company was a corporation based in Virginia, "with authorized capital stock of $20,000,000," and the intention of "a great development of power by means of hydroelectric machines to be located in caves excavated behind the falling water at the great cataract" (Adams, 1927). By using the falling water at the cataract site, they could avoid legal problems associated with diversion, not to mention the avoidance of much capital investment in an excavated canal or tunnel. Two local Buffalo businessmen actively supported this proposal "until convinced by a visit from Peter A. Porter, of Niagara Falls" that an alternative plan Porter was developing (the Niagara Falls Power Company) "was far more promising of practical success" (Adams, 1927). What Porter's plans were will be indicated below.

The advantages of having a distinguished reform-lawyer as president of the Commission is evident here in Green's clear understanding of the laws establishing the Reservation and the legislative processes of the Legislature. Note his specific reference to the "getting rid of mills and machinery, a part of which was an electric plant," the "Edson" (sic) electrical equipment on Bath Island (see Gardner report above) in the series of letters of protest sent by Green to various members of the Legislature. According to the subsequent and seventh annual report (1891), there was a public outcry against such use of the cataracts "to utilize the waterpower of the cataract for commercial purposes. The emphatic condemnation which that project received was significant...."

7th Annual Report of the Commissioners of the State Reservation at Niagara for the fiscal year from October 1, 1889 to September 30, 1890. Transmitted to the New York State Legislature January 29, **1891**.

In the seventh annual report (1891) for 1890, appeal was again made to the Legislature for adequate funds to get the Reservation along on its stated course. "A parsimonious policy towards Niagara, on the part of the State, should no longer prevail. There seems to be an impression in some quarters that the Reservation has proved to be an unwise and unprofitable investment, and somewhat in the nature of an expensive luxury." Green marshaled evidence that little expense is required of the Reservation. The Commissioners requested that it be "provided by law that a sum equal to the receipts from the Reservation ... be added to the sum annually appropriated for maintenance."

The Commissioners indicated their surprise that the Legislature would provide no money for "restoration of the Scenery of the Falls of Niagara," especially after the decade of effort to establish it. The attractions of Niagara were compared to those of the Yosemite Valley and Yellowstone National Park. Again the Gorge, with its "wild and picturesque scenery ... with the Whirlpool in the distance and the Rapids hastening, though with diminished speed, towards Lake Ontario" is included in the special environment at Niagara. "The pleasing hope may still be entertained and cherished that, at some day in the near future, the Whirlpool and the Rapids below, may be included within the public domain at Niagara."

Reference is first made here to a Niagara Falls Power Company, a company "composed of prominent business men at Niagara Falls" who have issued a prospectus to build a tunnel under the village of Niagara Falls, whose inlet would begin in the upper river, for the purposes of generating and transmitting power "by cable, pneumatic tube, or electricity" to associated companies. "The company has purchased about 1,300 acres of land near the Reservation" for mill-sites and homes for workmen.

The Commissioners direct the Legislature's attention to great national movements to preserve native American scenery, such as Yosemite and Yellowstone and the move to protect the remaining forests of New York State in the Adirondack Preserve, established in the same year as the Reservation, and other areas of distinctive scenery in the State.

Restoration of Niagara's vegetation is again referred to, the mainland shoreline to be "restored as nearly as may be to the condition it was in 200 years ago." By Chapter 570, Laws of 1889, the State Engineer and Surveyor was "authorized to make such surveys and maps and to set such monuments as may be requested of him by the said Commissioners."

Green reported that, because of his special effort, a "folding guide, including a map of the Reservation and suggestions to visitors" was issued.

At the end of the seventh report, John Bogart, State Engineer and Surveyor submitted a letter to Green "as to the Diversion of Water near Niagara Falls." There is a reference to the tunnel already under construction and that Mr. Bogart was in the village to make his own observations. In his opinion "the effect of the water flowing into this canal will ... be distributed over the whole river, and will not at all be confined to one section of it." In 1868, the volume of water flowing over Niagara Falls was measured by the Army Corps of Engineers "in connection with the survey of the Great Lakes. In 1868 it was known that the flow was 275,000 average cubic feet per second, or 16,500,000 cubic feet per minute. The proposed canal of the Niagara Falls Power Company was intended to divert 10,000 cubic feet per second or 3.64 percent of the total volume of the river.

Mr. Bogart refers to an interesting "hydraulic law" which will have significance when determining the visual impact of water diversion: that water directly over the crest of the falls has a greater velocity than in the river above or below it. As a consequence of this greater velocity, the depth of the water at the crestline is naturally less than it would be. Decrease in volume in the upper river due to diversion might exacerbate this effect. "But in fact, the decreased volume will give a decreased velocity, and therefore a greater relative depth at the crest."

Mr. Bogart concludes that "it is my opinion that the amount of water that can be taken through this tunnel will not affect the depth of water flowing over the Falls to an extent that will be visible."

This may have been true for the canal being built by the Niagara Falls Power Company, but did not take into account that of the hydraulic canal of Mr. Schoellkopf then in operation, with a history of expansion, nor of the growing number of business interests that would soon bid for their own diversions, and the ultimate effect this would have on the water levels at the falls. The water was and is not uniformly distributed across the crest. On the flanks of the falls the water was very shallow, particularly on the flanks of the Horseshoe Falls. Diversion would expose the bedrock of the flanking shorelines on the mainland shores of New York and Ontario, and the shores of Goat Island.

8th Annual Report of the Commissioners of the State Reservation at Niagara for the fiscal year from October 1, 1890 to September 30, 1891. Transmitted to the New York State Legislature January 29, **1892**.

In the eighth annual report of the Commissioners (1892, for 1891), it was determined that bad roads and walks interfere with the pleasurable sensations so important to the establishment of the Reservation. Improvements of the roadways on Goat Island had not yet been authorized by the Legislature. Importance in placed on installing the road so as not to disturb old trees. These trees were imperatively valuable because they were the ancient trees which looked down upon the historic events of European settlement and Indian occupation. Note that none of these trees are present today on Goat Island, none of which appear today to be older than the Reservation itself.

Note that again the restoration has not proceeded, but when it does, it will be "the most important and delicate which the commissioners will have to accomplish. They will secure in the prosecution of it, the assistance of the highest talent in landscape architecture" and of necessity will rely on "'the restoring processes of time and nature'". Note that the condition of the mainland shoreline sections of the reservation, with its denuded banks, had been lying unplanted for several years. By 1892 "the appearance of the river bank along the Rapids could have been made very attractive and the desired restoration far advanced" were it not for the Legislature failing "to respond to [the Commissioners'] appeals and [ignoring] their recommendations". An extended quote is included by the Hon. John Ferguson of Colombo, Ceylon, describing the intense emotional effects due to characteristics of the aqueous landscape, the volumes of water.

In anticipation of diversions proposed by the Niagara Falls Power Company of the previous annual report of 1890 (6 Ann Rep Comm), a description is made of the discovery of the Grand Falls in Labrador by American explorers "this year," 1891 or 1892, and a comparison of the two great cataract systems. Immediately after, reference is made to a perceptible drop in the water level of Lake Erie, which directly feeds the volume of water in the Niagara River, or strait, together with a diminution in the volume of water passing over the falls. There were problems in 1891 at the Maid of the Mist Landing because the water was so low "that it has ... been difficult for the steamboat ... to effect a landing at the dock " in the lower river below Prospect Park. The Commissioners request the Legislature to be aware of any reasons that come to their attention as to the causes of low water in the Niagara strait, and "scrutinize with great care and even refuse to enact all bills the object of which is the utilization of the water power of the river above the falls for manufacturing and other purposes. The falls themselves being within the limits of the Reservation, are no doubt secure from successful attack; but hardly a session passes without the introduction of one or more bills in the interest of companies organized for the purpose of utilizing the water power of the Niagara river, with the sanction of the Legislature."

In this annual report, it occurs to the Commissioners that the boundaries of the Reservation were not adequate to protect against this unanticipated attack on the spectacle of the Falls. They note that almost immediately upon establishment of the Reservation, there was an upsurge in demands on the water, and the Legislature made no hesitation at all in a vote favoring the power interests against the park. In fact, the Legislature proceeded to effectively harass the administration, thwarting their objectives by denying them adequate money.

The Commissioners include a warning to the Legislature by making from some document the following quote: "The improvident granting of franchises of enormous value, without recompense to the State or municipality from which they proceed, and without proper protection of the public interests, is the most noticeable and flagrant evil of modern legislation." The source of this quote may derive from a federal court decision with respect to monopolistic practices then underdoing the beginnings of federal regulation, particularly those of the railroad companies.

9th Annual Report of the Commissioners of the State Reservation at Niagara for the fiscal year from October 1, 1891 to September 30, 1892. Transmitted to the New York State Legislature January 31, **1893**.

In the ninth annual report of the Commissioners (1893, for 1892), reference is made to the State bonds sold to create the Reservation:. "but $300,000 of the Niagara Reservation bonds remain outstanding, and that within three years these will have been retired". By 1892 there were 'sufficient moneys in the treasury with which to pay these bonds after meeting all the appropriations made by the legislature and now in force.'

Here came a request by the Commissioners to put in a road between the Reservation and the Whirlpool: "It is still to be regretted that the Whirlpool could not have been included within the limits of the State's property at Niagara. It is hardly necessary to say that the Whirlpool and the Whirlpool rapids are only less interesting and less impressive than the Falls themselves. No intelligent visitor should fail to see them." In the fifth annual report (1889) note was made of carriage services to the Whirlpool and Whirlpool Rapids "where admission fees are charged and commissions paid."

Again a request is made to the Legislature to "refuse to grant to individuals or corporations the right to divert the water of the river for manufacturing or other purposes"

10th Annual Report of the Commissioners of the State Reservation at Niagara for the fiscal year from October 1, 1892 to September 30, 1893. Transmitted to the New York State Legislature January 28, **1894**.

In the tenth annual report (1894, for 1893), the Commissioners relate that the Niagara Falls Hydraulic and Power Company, operator of the hydraulic canal then in operation, "sought to procure the enactment of a bill giving it permission still further to divert the water of the Niagara river, and to enlarge its existing canals for this purpose." Green personally appeared before the "Assembly committee of commerce and navigation in the interest of the State against action by the committee favorable to the bill. He was ably assisted by the Attorney-General. The bill was not reported." The Commissioners "would again record their strenuous and unchangeable opposition to all proposed legislation of this nature and to all schemes or enterprises in the interest of corporations seeking to obtain the right to divert the water of the upper Niagara."

There was a lack of visitors, travel curtailed due to "disturbances in the financial world and the stringency in the money market, which were not favorable to traveling for pleasure." The Cataract Bank at Niagara Falls, for example, became insolvent June 23, 1893.

"The [State] Comptroller was authorized by law to cancel the remaining Niagara Reservation bonds held by the State and amounting to $300,000. The bonds were duly retired."

Note the order in which the Commissioners state their duties:

1. to protect "those portions of the Reservation that still preserved their original and distinctive natural character"

2. "to restore ... those portions that had suffered injury and defacement, to a condition that would be in harmony with those features of the scenery that had remained inviolate."

Only thirdly did the commissioners feel it was their duty "to provide for the conveniences, safety, pleasure and instruction of the public."

The Commissioners state that "a visit to Niagara should be an educational experience of genuine and permanent value" in contrast with stunts and spectacles.

The Commissioners now state that the Legislature should grant them the money to hire "the resources of landscape architects" who would be "tasked so as to produce natural effects that would leave the best and most enduring impressions upon visitors, and establish the most agreeable and perfect congruity between the grand spectacle of the falls and their surroundings." As much as possible, the restoration would be a duplication of natural processes for "nature herself did no more than this in the beginning, and all efforts should tend towards following her example and furthering her designs."

11th Annual Report of the Commissioners of the State Reservation at Niagara for the fiscal year from October 1, 1893 to September 30, 1894. Transmitted to the New York State Legislature January 28, **1895**.

In the next year, 1894, new bridges were built, including one from Goat to Luna Island (11 Ann Rep Comm, 1895). The Niagara Falls Hydraulic Power and Manufacturing Company "has been permitted to erect and maintain cribs and other structures in the Niagara river eastwardly from and outside of the reservation." Green opposed a bill to incorporate the Niagara, Lockport and Ontario Power Company, however "the bill incorporating the company passed the Legislature and was signed by the Governor" and the company "obtained permission to take water to an unlimited extent from the river, and to intercept and divert the flow of water from lands of riparian owners," the State being the riparian owner "at the Falls and for about a mile eastwardly from them."

Green went to the Convention which intended to revise the State Constitution in May of 1894, and had a committee established to report "whether an amendment should be made to the Constitution restraining the Legislature from granting to the corporations or individuals the right to divert the waters of the Upper Niagara...." An amendment was proposed by this committee that the Commissioners of the Niagara Reservation have direction and control of "all corporations, associations or persons who have ... been licensed or granted the right to divert the waters" of the Niagara River, including compensation to the State for such diversion. However, the Convention declined to pass any amendment "having for its objective the protection of the cataract."

Green insisted that the Convention "should at least have adopted an amendment providing for compensation to the State and limiting in some way the amount of water to be taken, and the purposes for which it might be used" such as sanitary purposes. Cholera and typhoid fever were to continue to be the scourge of municipalities along the Niagara River and elsewhere.

Green and his Commissioners in this report now came to realize that, although the State mandated their Commission, and the Province of Ontario the Commission of their park, and although it set aside the land adjacent to the Falls to protect its natural integrity by giving it "government protection," that the Legislature was showing no stint in permitting water power development.

This development was proceeding with no reference to the existence of the Niagara Reservation and its legislated mandate, and over the "strenuous opposition of the Commissioners and the people." The Legislature "granted to several corporations, without compensation, rights and licenses which, if valid, are of enormous value. One might think, from reading the daily and periodical press, that the Niagara river's principal use and function would henceforth be to turn immense turbine wheels ... in order that power may be developed to be sold by a great corporation; that the rapids and the Falls would hereafter be inferior in importance to the costly plant and structure of the same powerful corporation." These were oblique references to the Niagara Falls Power Company.

It is in this report that the Commissioners let it be known that a similar proliferation of hydropower development was occurring on the Canadian side, although the Commissioners of that property were able to exact "substantial annual rental or remuneration" for such privileges within the park boundaries. The American park received "no pecuniary compensation whatever."

The Superintendent declared that the mainland shores had been graded, filled and covered with sod or sown with grass seed. "The grounds on the mainland are now ready for the planting of such trees and shrubs as may be directed." This is a reference to the Olmsted-Vaux plan of 1887, indicating that a green lawn was only the preliminary part of the plan to restore the scenery. Note that the mainland has always stayed at this level of its restoration, except for the treatment of alien willow species, and horticultural shrubs as will be detailed below.

It is apparent that Superintendent Welch was very active in following the 1887 plan for "Gravel walks according to the plan have been constructed in the territory lately graded, extending along the rapids from Bridge street to First street." All through the various debates which involved the Commissioners, Welch was assiduously improving the Reservation under his care.

Details of the power companies presented in the report of the subcommittee of the Committee on Legislative Powers referred to above are as follows (Document No. 60, 1894, 11 Ann Rep Comm, 1895).

1.

The Niagara Falls Hydraulic Power and Manufacturing Company, organized around 1879, with no special rights or charter from the State. The company "is now engaged in increasing the capacity of its canal to 100 feet in width and fourteen feet in depth."

The company pays nothing to the State for its privilege.

Note that this company had been serving the electric needs of the Niagara Reservation from its inception, and for which the State paid a utility bills (see disbursement lists published in the annual reports of the Niagara Commission).

2.

The Niagara River Hydraulic Tunnel Power and Sewer Company of Niagara Falls, granted a charter in 1886 by the Legislature for building a hydraulic tunnel or sewer "for public use of sewerage and drainage, and for furnishing hydraulic power in the town of Niagara, Niagara county." Further increases in the powers granted to this company were made by the Legislature again in 1886, 1889, 1891, 1892 and 1893. This company changed its name to the Niagara Falls Power Company in 1889.

This company "has been given extraordinary and almost unlimited power in different directions. It has the right to condemn land; it may increase its capital stock to $10,000,000; its rents or charges for electricity or water ... are made a lien upon the premises on which such water or electricity is used ... it is authorized ... to conduct convey and furnish the waters of the Niagara river ... and ... said corporation ... may enter upon any private property for which it may obtain such right, or upon any public bridge or street ... and may use the ground thereunder ....'"

"It has paid nothing for the privileges granted, and has not agreed to pay anything, except that by one of the amendments to its charter ..." that it was to furnish, without charge, power and electricity and water for use in the Niagara Reservation.

"Your committee understands that the Commissioners of the State Reservation have never called upon this company to furnish either power or light, but have insisted upon paying for whatever the company might furnish them." Note that this provision will be referred to many times in the years ahead. Use of the advantages of the provision would be resisted until 1898 when officers of this power company became officers of the Commission of the Niagara Reservation, after which use of this power would be aggressively pursued by the Commission, with successful demands to the Legislature for finances to purchase and install the electrical equipment necessary to utilize the free power available by the provision.

Green was quick to see a bid for dependence by the Reservation on the power company written into this law, lured into tacitly accepting the company's diversion of water so as to receive lighting on the grounds.

Introduction of a legal provision to provide "free power" to the Reservation also laid the foundation for competitively excluding the company's rival, the Niagara Falls Hydraulic and Manufacturing Company which had been selling the Reservation electricity, as noted above.

The Niagara Falls Power Company was authorized to divert up to six percent of the volume computed as flowing in the Niagara River.

"It may be said in passing, that the same company has obtained from the Canadian government a grant for the use of an equal amount of water to be taken from the other side of the river", and so, twelve percent of the volume of water in the Niagara River.

3.

The Lockport Water Supply Company received a grant by the Legislature in 1886 "to take water from the Niagara river at a point near Tonawanda." This company became the Lockport Water and Electric Company since 1890. No limitations were imposed on this company with respect to the amount of water authorized for diversion, nor any compensation given to the State in return for "the powers and privileges granted."

4.

The Niagara County Irrigation and Water Supply Company incorporated in 1891 with the power to construct a canal with the Niagara River for manufacturing purposes in the towns of Niagara, Lewiston and Porter. No limit imposed on the amount of water to be diverted, nor any compensation demanded.

5.

The Lewiston Water Supply Company was incorporated in 1888, with no diversion limit imposed nor compensation required.

This company failed to achieve its aims and lost its New York State charter in 1893, although prior to this date, its entire capital stock provision was purchased by agents of the Niagara Falls Power Company in February, 1890 (Adams, 1927, Vol. 1:93). It was probably a failed venture due to technological limitations at the center of its design: the "mill over a wheel-pit—a method that received its knell at Niagara in 1890" (Adams, 1927, Vol. 1:93). Again, it appears as though representatives of what would be or was the Niagara Falls Power Company were able to forestall development of this company, as they were the Niagara Hydraulic Electric Company of 1887, the latter by the persuasion of Peter A. Porter, Jr., a Niagara Falls Power Company director (Adams, 1927, Vol. 1:96).

6.

The Buffalo and Niagara Power and Drainage Company was incorporated in 1889, authorized "to build a canal from within the city of Buffalo to a point below Niagara Falls," without diversion limits.

7.

The Model Town Company was incorporated in 1893, its name changing to Niagara Power and Development Company, to take water from the Niagara River at LaSalle in a canal. No diversion limits were imposed, and no compensation made to the State in return for its privileges.

Adams (1927) described this company, entitled "The Model City," which was to develop land along Lake Ontario up the Niagara River to Lewiston for sale to industries and the workmen's communities associated with them. There was to be a power canal "to take water from the Niagara River, east of the town of La Salle ... providing power by a fall at the escarpment ...," (Adams, 1927). "The company did some work of construction upon the sections of its canal at the proposed inlet and the outlet at the escarpment. The companies became bankrupt, and the moneys invested in the enterprise were lost" (Adams, 1927). The abandoned canal east of La Salle may be what we now know to be the "Love Canal."

8.

The Niagara, Lockport and Ontario Power Company was incorporated in 1894 to divert water from the Niagara River by canal "to the escarpment at or near Lockport" with no diversion limit, "and it has paid nothing, and has not promised to pay anything, for its rights and privileges.

This company, "incorporated May 20, 1894, to engage in the production and distribution of electric power in western and central New York State" (Adams, 1927, Vol.1:352-353) would later own and operate a hydroelectric plant on the Salmon River near Syracuse, a plant on the Oswego River, etc., and by around 1926, bought 130,000 horse-power from the Niagara Falls Power Company and affiliates.

One can assume that this company later built the Ontario Power Company generating plant at the base of the Horseshoe Falls—they do not appear to have had a plant at the cataracts in American territory. According to Van Cleve (1903), the Ontario Power Company plant was an enterprise "in which Buffalo capital is so largely interested." In their history of Buffalo, New York, Brown and Watson (1982) refer to John Joseph Albright as "one of the moving forces behind the Pan-American Exposition" who "helped form the Lackawanna Steel Company, interested himself in the New York Central Railroad system ... and assisted in harnessing the power of Niagara Falls by means of the Niagara, Lockport, and Ontario Power Company." By 1917, this plant had been acquired by the Hydro-Electric Power Commission of Canada (Ontario-Hydro), (Adams, 1927).

All of these companies "have received extensive powers and privileges, and their operations can be carried on almost without limit."

The subcommittee of the Committee on Legislative Powers warned that the impact of these diversions would affect the American Falls, due to its shallowness and consequent low volume compared to that descending over the Canadian Falls.

"In return for these immensely valuable franchises, the State of New York has not received one cent or any consideration, except the condition by the Niagara Falls Power Company to furnish the Niagara Reservation with water and light. While your committee understands that for a grant similar to the one obtained on this side of the river, this same company has agreed to pay our thrifty Canadian neighbors several thousand dollars per year rent, viz., about $25,000, depending upon the amount of water used."

The subcommittee goes on to say that by 1894, the citizens of New York State had paid $2,500,000 for the Niagara Reservation, but "if corporate and individual ambition is not checked and made subordinate to public rights, there is certainly danger that the Falls of Niagara ... will be sadly deficient in the amount of water flowing over their brink."

The subcommittee is worried that were all these concerns to divert any major extent of the (unlimited) amount legally granted them, there would be serious loss of water volume at the cataracts.

The subcommittee recommended that the above listed incorporated companies be placed under the control of the Commissioners of the Niagara Reservation, which is "charged with the duty of protecting, preserving and caring for the property of the State at the falls." The Commissioners would then balance the aesthetic requirements of the cataracts, and "the manufacturing interests which may inure to the benefit of the western counties."

As noted above, this recommendation was rejected by the State Committee on Legislative Powers.

In the twelfth annual report (1896, for 1895) the attitude of the people of western New York State as spoken for by the subcommittee of the Committee on Legislative Powers was described as follows:

"It is doubtless true that the people in the extreme western part of the State, to whom the sight of the Falls and river is a stale and almost daily exhibition, view with satisfaction the increased prosperity which may come to their particular section as the result of the operation of these various companies, and look with disfavor upon any attempt to preserve the property of the State, which may, at the same time, curtail their individual success."

Note that this characterization of "the people in the extreme western part of the State" contradicts statements made during the campaign to have the Niagara Reservation established where such a plan appears to have been welcomed by these same people. It is apparent that there were two distinct constituencies mutually opposed in this region. Naturally, those opposed to the Reservation were the developers—but who were the advocates? According to Welch (1903) the advocates were also the developers. They were advocates during the 1880's, and opponents during the 1890's. Some attempt to reconcile this apparent contradiction will be attempted below when the movement to "free Niagara" is reconsidered.

It is also apparent that if protection of the falls after the 1880's could not be provided by State government in Albany or by local municipalities in western New York State, then it had to be sought at the federal level of government. The values embodied in the cataracts as put forward by Olmstead and others had national and international significance, but the hydraulic and hydroelectric values to local and regional industry were promoted and developed without recognition of the concerns of the American people in this regard.

12th Annual Report of the Commissioners of the State Reservation at Niagara for the fiscal year from October 1, 1894 to September 30, 1895. Transmitted to the New York State Legislature January 28, **1896**.

In the 12th Annual Report of the Commissioners (1896, for 1895), Green explored further the legal implications on the federal level for controlling excessive and destructive diversions from the upper Niagara River, as suggested in the report of the subcommittee of the previous year.

In the first paragraph, Green stated his confusion, and that of others who had promoted establishment of the Reservation, as to why the State should, so soon after establishing the Reservation, work so industriously to destroy it. The Commission opposed incursions on Reservation lands: "The lands of the Reservation were to be restored to their natural condition. And although corporations and individuals have sought to secure a foothold within the Reservation for private pecuniary profit, they have thus far always been defeated" (12 Ann Rep Comm, 1896).

"It probably did not occur to those who interested themselves in the work preliminary to the establishment of the Reservation, the work of arousing public sentiment and securing legislative approval of the project, that the State, which in 1885 was willing to expend a large sum of money to make Niagara free, would so soon tolerate, authorize and legalize schemes that could not but be injurious to the Reservation and in antagonism to the purpose of the State in establishing it."

The development rush — The reader's attention should be addressed as well to the fact that in all the six years it took to push the Niagara Reservation through the Legislature, no mention was ever made of the threat of power development, and yet how extraordinary was the rush to the Legislature, almost immediately after the Reservation was established in law, to claim rights now understood as necessities for the development of utilities. It is as if all these interests waited for the Reservation to be established, as if its establishment was crucial in some way, for the industrialization of the falls at Niagara to proceed.

The only instance where the "park people" who pushed for park legislation showed foresight in this respect (and whether this was intentional or not is difficult to say) was the assignment and appointment of Andrew H. Green, with his powerful association with reform politicians, his commitment to public works of a social character, and his distinguished legal background, to the office of President of the Niagara Commission.

Green time and again in his reports to the Legislature expressed his surprise and perplexity toward this body, which so recently bore in triumph the State's commitment to national, historical, natural and recreational values, and which was now concentrating on industrial development of the very landscape which was its honor to protect.

A proposal to the Legislature had been made in 1895 to authorize a company to construct a cable-car arrangement in the vicinity of the falls, stretching between the Canadian and American shores "to enable passengers to get views of the Falls and the river," much like today where all kinds of proposals are made by developers to erect structures which provide views so as to derive a profit from visitors, most notably being helicopter rides. Such bids for development are, according to Green, "too preposterous to merit serious consideration; nevertheless, [such proposals] may be brought forward again, and unless stoutly opposed, may yet become an accomplished fact."

In addition to the Legislature freely giving legal rights to divert water and develop the upper river, and to entertain private development at the falls and along the gorge for the personal profit of some, the Legislature was also examining proposals by private companies, in this case, the Niagara Falls and Suspension Bridge Railway Company, demanding "rights" deemed by themselves not extinguished during the condemnation proceedings during the 1880's, rights which would allow developers to develop within the boundaries of the Reservation itself.

It appears that developers in the area of Niagara Falls had a clear idea that the Legislature, far from taking a protective role at Niagara, was in fact the source for handing out legalized opportunities for commercially developing the falls and its gorge. The Legislature did not have this characteristic before the movement to preserve the falls, but immediately afterward, it did. As a matter of fact, the political mechanisms for establishing the Reservation revealed weaknesses in the State political system. This system was then in such deep need of revision that it formed committees for governmental reorganization. This system, by which the park legislation, first thought to stem degradation at the falls, was created in law, also revealed the political means for Niagara's continued degradation. Such degradation now, however, rather than involving such low-capitalized concerns as hotels, mills and such, involved highly-capitalized corporate concerns with great influence in Albany, which were capable of accumulating capital from ten to fifty millions of dollars.

John Jacob Astor — Director, Cataract Construction Company, director, Niagara Falls Power Company (Adams, 1927).

This was the age of great capital ventures, the "days of large enterprises" (Barhite, in 11 Ann Rep Comm, 1895), which could now be capitalized by corporate concerns controlled by such legendary American capitalist families as the Astors, Morgans, Vanderbilts, and so forth. The days when capital had to be gotten from Europe to capitalize great public works, such as the Erie Canal, were nearly over as far as the United States was concerned. This was not the case for the British territory of

Canada, which was far poorer in capital. An expanded tax base and a strong, evolved American private economy allowed Americans to go ahead with major developments using their own resources.

The electric utility is a kind of mongrel company operated by private concerns but given legal privileges by government yielding them the characteristics of a state-sponsored monopoly. Due to the great benefits derived by the people dependent on this kind of company, utilities were beginning to emerge across the United States in parallel with technological innovation. Such companies required an immense amount of capital to be developed and a long period before profits, commensurate with the amount of initial investment, could be realized. Governments unable or unwilling to accumulate this capital to develop such a source of energy, became legislatively lenient toward such developments on which the citizenry would ultimately become dependent.

Green repeated in the twelfth annual report (1896), as he did in all his annual reports as President of the Commission, the mandates to which the Reservation and its officers were committed: to restore and preserve the natural condition of the area around the falls, and to keep it free for public enjoyment.

Green warned the Legislature, as a body, that if it enacted laws "allowing private interests to obtain rights within the Reservation" thereby nullifying the act of 1885, "the opposition would be so strenuous as to prevent its eventual success."

Green stated that the series of grants given the private companies by the Legislature, outlined by the subcommittee of the previous year and listed above, "are indefensible and antagonistic to the public interest." Green stated that "it is not unreasonable to question the existence of the moral right to grant to private corporations, without compensation, what belong to the people."

As to the sister Commission on the Canadian park, they had granted the Niagara Falls Power Company the right to divert water on the Canadian side, and to build a tunnel and a plant within the park boundaries. Green could not tell what the Canadian people thought of this concession.

It was also known to Green that those objecting to a State Constitutional amendment to permit the Niagara Commissioners to regulate the diversion of water from the upper river declared that such an amendment would be useless, "since the corporations would be able to secure the right or license to use the water from the Ontario or Canadian authorities, and transfer their plants and bases of operations to that side of the river, with the consequent pecuniary and other material advantages"

(Green, letter to the Secretary of State, Washington, 12 Ann Rep Comm, 1896).

Apparently, those representing the American corporations in this matter already knew that the Commissioners of the Canadian park would not hesitate to attract development into the Niagara park established in that country. Even though the Canadian Commission exacted fees for use of their resources developed in their park, there were still "pecuniary and other material advantages" to putting their plants there. Indeed, one might imagine that Commission doing their best to attract development.

In the twelfth report, Green already acknowledged the failure of the Provincial and State governments to protect the falls, and questioned "whether protection cannot be insured by the intervention of the governments of the two nations whose territories are bounded by these waters." Green recognized the weakness of the governments of the province and state, and recourse would have to be taken on the national level to achieve truly disinterested protective legislation, again: "If the State of New York and the Province of Ontario fail to interpose for the protection of the great cataract from threatened injury, the question arises whether protection cannot be insured by the intervention of the governments of the two nations whose territories are bounded by these waters."

The federal government has jurisdiction over the waters of the Niagara River because the river, in spite of the cataracts, was deemed a public navigable river, and the international boundary passes through it. By right of sovereignty, the national government of the United States could engage with Great Britain in treaty negotiations to control issues relating to the preservation of Niagara Falls. The State of New York "it need hardly be said, cannot enter into negotiations with a foreign power."

Green had no reservations about the ability of the New York State and Ontario Provincial Legislature to uphold their legislation relating to Niagara Falls: "If the United States and Great Britain refuse to interpose, there is nothing to prevent the State of New York and the Province of Ontario from drawing off so much of the water of the upper Niagara as to make the cataract practically disappear."

It is also apparent that Green did not yet consider the possibility of the federal government interposing at Niagara as it did in Yosemite, making the falls and gorge of Niagara a national park.

Natural grandeur — New York State was the custodian of a resource of international concern: "that the Falls of Niagara shall be preserved in all their natural grandeur, in order that men of all nations may resort thither for edification and delight henceforth and forever."

In that year, Samuel Parsons, Jr., was appointed "landscape architect to the board," and according to Parsons' direction, the banks of the river were graded and planted "so as to fringe the banks in a natural and picturesque manner." "Native specimens have been chiefly used, and the aim has been to produce a wild, natural effect, in harmony with the surrounding scenery" according to the conceptual framework of Frederick Law Olmsted and six years of legislated intent. Parsons had visited the Reservation on April 29.

Although not part of the plan of 1887, Thomas Welch, Superintendent, continued his construction of a series of stone bridges, rustic in appearance, on the mainland shore, associated with the various channels left behind from the races built by former mills. He proceeded vigorously to improve the quality of the buildings, the services and the implementation of the 1887 plan.

At the end of the twelfth report are a series of letters Green wrote in 1894 to Walter Q. Gresham, Secretary of State, Washington, D.C., J. W. Langmuir, Esq., Chairman Commissioners Queen Victoria Niagara Falls Park, and Theodore E. Hancock, Attorney-General of the State of New York. Gresham was asked to open correspondence between the United States Department of State and the Canadian Minister of State to discuss regulation of diversions and other issues relevant to the preservation of Niagara. Langmuir was asked to cooperate in securing "an international agreement, to the effect that hereafter no right or license to divert the water of the river shall be granted to any corporation on either side of the river," such diversion being considered "the destructive designs of corporations organized for mere money-making."

The Niagara Falls Hydraulic Power and Manufacturing Company was widening the hydraulic canal, and Hancock was asked whether this company had acquired the legal right to divert water from the river, which belonged to the people of the State of New York. In Mr. Hancock's legal opinion, the company had been granted no right to expand its canal, nor to divert waters for power purposes "without a grant or prescriptive right." Its expansion exceeded the rights granted the company "under the act of 1875, chapter 611." Mr. Hancock distinguished clearly between two uses to which the water may be put by people diverting it: water used for domestic purposes, and water used to turn a profit. He also presents legal objections to constructions in the river which impede the right to navigate the river, and the interpretation of diversions technically as nuisances in law, and subject to restraint the by Commissioners of the Niagara Reservation. Interested readers are directed to this letter of Mr.

Hancock who presents technical legal information to accompany his determinations.

The Commissioners were each made by appointment. It is easy to see how potentially powerful the Presidency and Commissioner's seats were on this board in determining the regulation of the diversions and other developmental issues with respect to private industry—powerless if one were attempting to regulate the power companies, influential if one wished to use these offices to promote power interests.

With respect to this power of the Commission, the reader's attention is directed to one of the proposals up before the Legislature in 1895: "Another scheme last winter was to permit a tramway company to run its tracks into the Reservation. A bill was introduced in the Senate entitled, 'An act to confirm the right of way of the Niagara Falls and Suspension Bridge Railway Company.'" The attitude of Green and the Commissioners was firm: "It is hardly necessary to repeat that the commissioners are strenuously opposed to all schemes that are in any way intended to permit private persons or corporations to obtain rights within, or to encroach upon the Reservation.... It would seem to be the duty of the commissioners to take, if possible, a more determined attitude than ever in opposition to all measures of this sort."

Later, in 1898, when the members of the Commission were changed, but for Green and Raines, this railway company was given its wishes by the new Commission, with consequences for the Niagara Reservation. This was the company referred to above, which claimed to still have rights to the properties now within the boundaries of the Reservation.

13th Annual Report of the Commissioners of the State Reservation at Niagara for the fiscal year from October 1, 1895 to September 30, 1896. Transmitted to the New York State Legislature February 1, **1897**.

The next year (1896), the Commissioners continue to urge the Legislature to protect the Reservation "from the insidious designs of corporations and individuals who seek to secure from the Legislature and from the Commissioners rights and privileges at Niagara in direct contravention of the purpose of the State in establishing the Reservation, and in flagrant opposition to the public interests" (13 Ann Rep Comm, 1897, for 1896).

The fact that Schoellkopf's hydraulic canal (the Niagara Falls Hydraulic Power and Manufacturing Company) successfully used Niagara water for manufacturing purposes, and the successful development of electric power by the Niagara Falls Power Company has "naturally directed the eyes of shrewd and speculative men towards a similar use of the river."

The Legislature hastened to address the legal vulnerability of the Niagara Falls Hydraulic Power and Manufacturing Company with respect to its expansion and diversion of waters outlined by the Attorney-General in the previous year, and granted this company the right to develop its canal, divert its waters and sell power developed from the diverted water for profit.

"The Commissioners are of the opinion that the Legislature should not grant to a private corporation, without compensation, that which belongs to the people."

All while urging the Legislature to restrain diversions from "certain speculative and manufacturing concerns in Niagara county [who] desire to secure enormous profits from the use of the waters of the Niagara river" the Commissioners took pains to reiterate the issues expressed in the report on the preservation of the scenery of 1880 by Gardner, Olmsted and the Commission appointed at the time, the literature generated in the intervening years and the legislation of 1885 establishing the Reservation. They include lovely pictures of Niagara's unique landscape and append essays of scientific importance. Increasingly, lovely photographs are appended of the new landscape created under Superintendent Thomas Welch for the mainland section of the Reservation, moonlit scenes of adolescent trees, grassy water verges, rustic stone bridges, particularly around what was known as Willow Island.

Rip-rap on the mainland shore and the periphery of Bath Island was derived, ironically, from "very large stones obtained from the excavation on the lands of the Niagara Falls Hydraulic Power and Manufacturing Company." The excavation was from the expansion of the canal just referred to by Green in this and the previous year. More canal material was taken to fill the pier at Port Day and extend it. Welch installed 625 feet of iron railing designed by Calvert Vaux along the west end of Goat Island.

The office building maintained on Bath Island was called by Welch the "office of the Commissioners."

14th Annual Report of the Commissioners of the State Reservation at Niagara for the fiscal year from October 1, 1896 to September 30, 1897. Transmitted to the New York State Legislature January 25, **1898**.

In the 14th annual report for the year 1897 (1898), among those indicated as officers of the Commission, Samuel Parsons, Jr. was listed as the official Landscape Architect to the Niagara Reservation. He visited the Reservation in June of 1897 to restore the area "between First street and the upper terminus of the Reservation at Port Day" along the mainland shore.

In this report the achievements of the year are recorded, made "in accordance with their previously adopted plan," that is, the plan of 1887. It is acknowledged that money for improvements, rather than simply maintenance, had been granted them only since 1889.

Andrew Haswell Green and power development — Green's battle with the power companies resumed, with a quote, the author of which Green declined to identify except to say an "eminent man of science." This individual told the press: "The originators of the works so far carried out and now in progress hold concessions for the development of 450,000 horsepower form the Niagara River. I do not myself believe and [sic] such limit will bind the use of this great natural gift, and I look forward to the time when the whole water from Lake Erie will find its way to the lower level of Lake Ontario through machinery doing more good for the world, than that great benefit which we now possess in contemplation of the splendid scene which we have presented before us at the present time by the waterfall of Niagara. I wish I could live to see this grand development" (14 Ann Rep Comm, 1898, for 1897).

The quote was of William Thompson, otherwise known as Baron Kelvin of Largs, or, Lord Kelvin. It was made during a visit to Niagara in August, 1897, and his last word was actually "I do not hope that our children's children will ever see the Niagara cataract" (Thompson, 1910). He was less visiting the cataracts than he was the power industry development in the vicinity.

Kelvin was a distinguished mathematician and physicist at the University of Glasgow. We owe to him the Kelvin scale for absolute temperature. "In thermodynamics his work of coordinating the theories of heat held by various leading scientists of this time established firmly the law of the conservation of energy as proposed by Joule" (Bridgwater and Kurtz, 1963).

International Niagara Commission, London 1890–1891 — Center: Sir William Thompson, Glascow. Ringing him, left to right Prof. E. Mascart, Paris; Prof. W. C. Unwin, London; Dr. Coleman Sellers, Philadelphia; and Col. Theo. Turrettini, Geneva (Adams, 1927).

In 1890, Sir William Thompson had accepted the position as president of a body called the International Niagara Commission sponsored by the Cataract Construction Company—the executive and financial body of the Niagara Falls Power Company. The commission "was to conduct a scientific symposium on the development of power at Niagara Falls, which would attract the best scientific and engineering knowledge and experience of those most competent to be found in the nations of the world" (Adams, 1927). His contempt for the preservation of Niagara Falls as a natural phenomenon and the public lands established for this purpose is a consequence of his interest in the generation and transmission of electrical energy for commercial ends. Green appears to have been dismayed that Thompson should have lent his international prestige to the destruction of the landscape under Green's protection.

Reference is made, in this Annual Report, to a growing movement in the State of New York to protect other areas of natural significance—the Palisades outside of New York City from quarrying, and the Adirondack forest preserve from logging. The State had incorporated The Trustees of Scenic and Historic Places and Objects to address the protection of the State's natural and cultural resources.

The Commissioners refer to the curious legal contradictions of the Legislature in depriving the pre-1885 owners of their profitable land now within Reservation boundaries in order to preserve the falls, and then giving way to other, larger corporations: "only a few years afterward ... by legislative acts [the Legislature] grants to corporations, without compensation, the privilege of taking enormous quantities of water from the river." In addition, by 1898, "an attempt may soon be made to divert large quantities of the upper river into Canadian territory."

The Commissioners ask "has the interest in the development of electrical power and manufacturing enterprises become so general and so dominant as to produce indifference to the future of Niagara Falls ...?"

15th Annual Report of the Commissioners of the State Reservation at Niagara for the fiscal year from October 1, 1897, to September 30, 1898. Transmitted to the New York State Legislature February 1, **1899**.

In the fifteenth annual report (1899, for 1898), the question was asked whether the nearby City of Niagara Falls, New York, was expanding into "one of the most conspicuous and important manufacturing centers in the country." This question appears to be really based only on the boasting of the power company in their advertising and promotional literature of the time that the community would become so—it never actually achieved such importance. The Commissioners emphasized that it "should not be forgotten that the Reservation really belongs to the State, to the whole State and not to any portion or section of it" and that "local interest in the Reservation is entirely subordinate to the interest of the State."

Events related in this annual report will be discussed in the sixteenth report below.

16th Annual Report of the Commissioners of the State Reservation at Niagara for the fiscal year from October 1, 1898, to September 30, 1899. Transmitted to the New York State Legislature February 1, **1900**.

In the year 1898, three of the Commissioners were changed (16 Ann Rep Comm, 1900, for 1899). On March 25, 1898, the Governor of New York State appointed Thomas P. Kingsford (Oswego), Charles Dow (Jamestown) and Alexander J. Porter (Niagara Falls) Commissioners of the Niagara Reservation, retaining George Raines of Rochester. Charles M. Dow later was to become Green's successor as President of the Commission after 1903. The Commissioners, according to Chapter 336 of the laws of 1883 would hold office for five years (19 Ann Rep Comm, 1903).

Henry E. Gregory of New York City, who had served as Secretary-Treasurer (Dec. 29, 1887 to Feb. 1, 1899) and whose name appears in the fifteenth annual report, had resigned and a new Secretary-Treasurer, Mr. Richard F. Rankine, of Niagara Falls, was elected in his stead. It was said that Gregory "presented his resignation November 21, 1898, and was succeeded February 1, 1899, by Mr. Richard F. Rankine" (16 Ann Rep Comm, 1900, for 1899).

The Daily Cataract reported the annual meeting of the new Commissioners in the Commissioners' office on Bath Island on May 19, 1898 ("New Commissioners meet here.") Note that this was the year prior to events reported in the 16th annual report. In this note in the Daily Cataract Green is referred to as "ex-President" suggesting that the Presidency was a matter of vote upon (re)appointment of a new Commission, and that the paper fully anticipated Green's dismissal as President. If that was the case, then the new Commissioners reelected Green President of the Commission anyway. All the newly appointed Commissioners showed up at this meeting. According to the paper, George Raines, the only other Commissioner to "hold-over" from the previous Commission (appointed March 22, 1893) was not present.

An about-face — There is, strikingly, no criticism of the power companies in the 1899 annual report, in fact, the free use of stone derived from expansion of Schoellkopf's hydraulic canal is noted with uncharacteristic gratitude: "The enlargement of the hydraulic canal, adjacent to the Reservation, affords a fortunate opportunity for obtaining the large stones used in riprapping the river shore, without expense except for hauling." This is the same canal the expansion of which Green had vigorously opposed a few years previous. The Commissioners also now want to take advantage of the free water and electric power granted the Reservation by law (Chapter 513 of the Laws of 1892) to operate the inclined railway at Prospect Point and illuminate the grounds and buildings, and for "sprinkling the lawns," although previously such a concession was looked upon with suspicion. It is in this year that the name of Bath Island is changed to Green Island, in honor of A. H. Green. It is curious that Welch uses the new name in the report for the previous year (1899), mentioning the removal of buildings, including the office of the Commissioners, from "Green" Island, before the new name had been made official.

When, in 1898, a committee was made of three of the Commissioners, Dow, Kingsford and Raines (Alexander Porter, the Commissioner not mentioned in the dispute, apparently abstaining), to examine a petition for the building of a street surface railway "upon and along that part of the Riverway ... between Falls and Niagara Streets," Green was shown as dissenting from the committee's decision to approve. This is the first indication that Green was losing control of the Commission. Apparently Green disapproved so strongly that he attempted to challenge the railway decision on legal grounds. The other Commissioners apparently sought a legal opinion from the Hon. Frank W. Stevens, who upheld the decision of approval (16 Ann Rep Comm, 1900, for 1899), against Green.

A report by the Daily Cataract describing the Commission meeting on Bath Island, mentioned above, for May 19, 1898, contradicts the later account presented to the Legislature over a year later. This committee was called an "executive committee" by the paper, and it was composed of "Mssrs. Dow, Kingsford and Porter" with no reference to Raines, who was said to be absent.

The railroad — Furthermore, the establishment of the railroad on Reservation land seems to have been the only order of business conducted by the Commission in that year. After establishing the "executive committee" composed of everyone but Green and the absent Raines, the meeting was adjourned.

On the next day the Daily Cataract reported (May 20, 1898, p.4, headline: "Little business") that "no appointments were made and no changes were ordered in the administration of the affairs of the reservation." The Superintendent's report was published in that journal.

The issue of the petition to the Legislature by the Niagara Falls and Suspension Bridge Railway Company, first came to Green's attention in 1895, which he and the Commission vigorously opposed (see above, 12 Ann Rep Com, 1896).

An extensive quote from a previous annual report made by Green, repeating the mandate to preserve the Falls from diversion, is included, giving the impression that the author of the annual report had little personal experience or conviction toward those issues, such that he had to quote from a previous publication. *It is doubtful that Green wrote the report*, and that the Commission now found itself without a spokesperson—temporarily.

The next year, 1899, the Commission met in New York City late in the year: December. During the middle of the year, Mr. Henry E. Gregory, of New York City, ceased to be Secretary-Treasurer as of February 1, 1899 (19 Ann Rep Comm, 1903). Mr. Rankine picked up the post, officially serving between February first of 1899 and February first of 1900, yet he handed in his resignation May 8, 1899.

The Niagara Falls Gazette made note of Mr. Rankine's resignation from his post as Secretary-Treasurer (May 8, 1899, p. 5) when it was reported that Richard F. Rankine "handed in his resignation on the day he was made secretary and treasurer of the International Traction Company. There is somewhat of a scramble for his position, which has a good salary attached. It is likely it will go to a well-known resident of Niagara Falls." All of the previous Secretary-Treasurers of the Commission, except for a brief one-year stint by David Gray of Buffalo, New York, in 1887, had derived from New York City. The International Traction Company was managed by the railroad company who would be putting the new railway into the Reservation (see below). When Rankine's successor was named, the Gazette repeated (Dec. 8, 1899, p.1) that Rankine resigned "to assume important duties for the International Traction Company." By 1901, that company, said to be of Buffalo, New York, would use "the largest utilization of water power for street railway purposes in the world," and that "Practically all this system is now operated by electric power derived from the power plant of the Niagara Falls Power Company" (General Electric Company, 1901).

When the Commission met in New York in December, the Niagara Falls Gazette reported the election of Peter A. Porter, Jr. as Secretary-Treasurer, although officially (19 Ann Rep Comm, 1903) he was said to have started later, February 1, 1900 (see Gazette, Dec. 8, 1899, p.1).

Peter A. Porter, Jr., in addition to writing a number of histories pertaining to events in the Niagara region, and published in the annual reports of the Commissioners, had owned the Niagara Falls Gazette after purchasing it toward the end of 1880, which may account for some of the Reservation's disbursements to Porter for printing charges: $14.50, 4 Ann Rep Comm, 1888; $36.25 for printing and advertising, 3 Ann Rep Comm, 1887. In later years, payments were made to the Gazette Publishing Company for the Superintendent's office expenses, probably pertaining to public announcements for Reservation business, such as minutes of Commission meetings, announcements of contracts, etc.

Mizer (1981) indicated that in the era of the 1880's and '90's many newspapers "were little more than 'personal journals,' dedicated to promoting the interests, usually political, of the editor and/or publisher." Porter, however, "throughout his years of ownership was devoted to constant improvement of the paper. Within a week after it became a regular daily, the Gazette was carrying Associated Press and United Press dispatches, giving readers good coverage of world events as well as improving local news coverage" (Mizer, 1981). P. A. Porter, Jr. sold the Gazette in 1895 to "a corporation headed by William B. Rankine" (Mizer, 1981), an executive of the Cataract Construction Company, director of the Niagara Falls Power Company (Adams, 1927).

Green's loss of control — A new issue regarding water volume in the Niagara River had arisen in 1899: "A bill has been introduced in the Congress of the United States providing for the construction of a dam or jetties at the head of the Niagara River, at Lake Erie, in order to raise the level of the water in the Great Lakes." Such a construction would have a seriously adverse effect on the amount of water available for operation of the manufacturing and electric companies. This issue was extensively discussed in the Niagara Falls Gazette (see, for example, December 11, 1899). The Commissioners state that "such an obstruction would be liable to diminish the volume of water flowing over the falls, and thus injure the beauty of the natural scenery of Niagara, which the State of New York has expended its means and exerted its authority to protect ...." It is deeply ironic that the authors of this protest represented the interests of the power companies who would divert so much water that federal legislation would have to be drafted to restrain them and their Canadian affiliates (see Taft Committee below).

Now the issue of diversion had taken on a national perspective, not a simple wrangle between Green and local developers over diversion and the destruction of the scenery at Niagara Falls. Such a structure at the head of the Niagara River at Buffalo would not only affect the scenery at the cataracts, but it would have a harmful effect on the power companies operating there as well, limiting their "limitless" potential for diversion granted by the New York State Legislature.

Perhaps this had something to do with the appointment of Alexander J. Porter to one of the posts as Commissioner, for Mr. Porter was a stockholder in the Cataract Construction Company, the holding company for the Niagara Falls Power Company, and an officer and director of the latter company under its 1886 name: the Niagara River Hydraulic Tunnel, Power and Sewer Company (Adams, 1927).

Green's campaign to appeal to Congress to control diversions injurious to the scenery of the Falls would now be taken over by power company interests who were also dependent on the volume of river water, and stood to benefit from these controls, at least where threatened by water use practices in the upper river and upper Great Lakes.

Further evidence of Green's loss of control of the Commission in this year and evidence for the beginning of the power company's taking control of the Commission begins with the passage in 1899 by the Legislature with the approval of the governor of an act (Chapter 710 of the laws of 1899) to amend a railroad law. The law was changed to grant exclusive control by the Commissioners of the Niagara Reservation for the construction of a street surface railroad on Reservation grounds and the issuing of licenses to use the tracks. Although Green was attempting to get the Legislature to extend the regulatory powers of the Commission so it could more effectively control issues potentially detrimental to the objectives of the Reservation, the Legislature had not cooperated. Yet in this year, by a special act, the power of the Commissioners to decide issues was suddenly given.

The new Commissioners, all but the President of the Commission, Mr. Green, taking advantage of the new legislation, and signed a contract with the Niagara Falls and Suspension Bridge Railway Company to "use the street railroad tracks upon and along that part of Riverway, so called, in the State Reservation at Niagara, between Falls and Niagara streets, in the city of Niagara Falls." The Niagara Falls and Suspension Bridge Railway Company, the International Traction Company, the Buffalo and Niagara Falls Electric Railway, Buffalo and Lockport Railway, the Niagara Falls Park and River Railway Company and the Buffalo Railway Company were all under the "same general management," and were granted rights in the contract just mentioned. Buffalo had built its first railroad in 1836, terminating in Niagara Falls. "Black Rock's Peter [B.] Porter was the force behind the Buffalo-Niagara Falls railroad

(Brown & Watson, 1982). The President of the Niagara Falls and Suspension Bridge Railway Company, W. Caryl Ely, of Buffalo, was an incorporator of and attorney to the 1886 company soon to become the Niagara Falls Power Company, of which he became a stockholder and trustee. According to the 1899 contract: "the motive power used" for the railway "shall be electricity exclusively," that is, not steam nor the horse-drawn type introduced into the City of Niagara Falls in 1883 (Scott & Scott, 1983, p. 40; 1882 according to Mizer, 1981).

The Niagara Falls Power Company incorporates — It is curious that Ely, former citizen and representative from Cooperstown, representing Otsego County in the Assembly, had been a strong supporter of the bill passed in 1885 to create the Niagara Reservation (Welch, 1903). Mr. Ely was a lawyer who, by February 3, 1886, had moved to Niagara Falls (Adams, 1927). As a matter of fact, he appears to have moved to that city immediately after passage of the Reservation bill in July of 1885, because he was representing Otsego County during its passage (Welch, 1903). According to Adams, Ely was a prominent member of the Democratic Party and "had participated in the final stages of the effort to make "Niagara free," and was a man of wide acquaintance and fully cognizant of the difficulties that might be expected to be encountered in any such matter requiring legislative action" (Adams, 1927). So expert was Mr. Ely in legislative matters that not six months had passed after the Reservation was opened with great fanfare before he had moved to Niagara Falls and drafted a bill before the Legislature, submitted to Peter A. Porter, Jr., the former "then member of the Assembly from the Niagara Falls district" who introduced it in the Assembly (Adams, 1927). By the end of March, 1886, the Legislature had passed the bill incorporating the Niagara River Hydraulic Tunnel, Power and Sewer Company of Niagara Falls, New York, which soon changed its name to the Niagara Falls Power Company.

ALEXANDER J. PORTER · PETER A. PORTER · MICHAEL RYAN

OFFICERS
AND DIRECTORS
OF THE
"GASKILL COMPANY"
THE
NIAGARA RIVER
HYDRAULIC TUNNEL,
POWER AND SEWER
COMPANY
OF
1886

· HENRY S. WARE · THOMAS V. WELCH

Officers and directors — Some of the officers and directors of the
Niagara River Hydraulic Tunnel, Power and Sewer Company, 1886,
including two Porters and Thomas Welch, discussed here. (From Adams, 1927.)

By 1899, Mr. Ely had become president of the railroad company, signator with the Commissioners of the Niagara Reservation, of
which one, Alexander J. Porter, had been appointed that year. As the
railway lines were reorganized under new names, Ely became president
of the International Railway Company and was the author of the plan to
merge the twin cities of Suspension Bridge and Niagara Falls in 1890. It
was subsequent to this date that Mr. Ely became president of the railway
company (Mizer, 1981). Mr. Ely was also president of the Buffalo and
Niagara Falls Electric Railway Company (Mizer, 1981), which may have
been an earlier name for the International Railway Company. According
to Seibel (1985), the International Railway Company came about during
a merger or company reorganization: "On April 18, 1899, control of the

Niagara Falls Park and River Railway was acquired by the Buffalo Railway Company which then became involved in a merger which produced the International Railway Company." "On February 12, 1902, the IRC [International Railway Company] formed merging 25 trolley, railroad, and bridge companies in Niagara County, Buffalo, and Tonawanda," (Scott & Scott, 1983, p. 56).

W. Caryl Ely

The plan for merger of the two towns was brought by Mr. Ely before the Business Men's Association, formed in 1890, of which he was presumably a member. It is this association which was referred to by Toucey, General Manager of the New York Central, in his dispute with Welch over the Miller & Co. carriage service, as will be discussed below.

Reference is made here to the increased population anticipated for the Pan-American Exposition, by Welch in the 14th annual report (1898, for 1897). According to him, the Exposition was to be held in 1899 on Cayuga Island, located just above (east) where the present North Grand Island Bridge is now. The anticipated populace was five million visitors who "will probably visit the Falls and other points within the Reservation" for which "preparation should be made in advance for the accommodation of a greatly increased number of visitors." The next year, a building in Prospect Point which had formerly held public gatherings at that spot, and was "accessible to visitors in case of sudden rainstorms" was burned down by an arsonist (Welch, in 15 Ann Rep Comm, 1899, for 1898). The next year, the Legislature had approved funds for the new building and the State Architect, not Vaux and Downing, was asked by the Commissioners to prepare the design, which was accepted and building commenced in the spring of 1900 (16 Ann Rep Com, 1900).

Niagara Falls Power Company fully operational — Peter A. Porter Jr., who would officially become secretary-treasurer of the Commission by the next annual report, and who was grandson of Gen. Peter B. Porter (see above), submitted and published in the sixteenth annual report (1900) his history of Goat Island. This is fifteen years after Goat Island became part of the Niagara Reservation, and by now the Niagara Falls Power Company, of which he was a director, was fully operational.

This essay beautifully relates the interesting character of the island's natural and social history, and, as intended, did much to establish the prestige of the Porter family who held the island as a family heirloom until 1885. Although acknowledging the family's industrial interests and development of the mainland at the cataracts—the very businesses removed in the establishment of the Niagara Reservation—Porter wrote: "The owners of the Island were then power users and power developers, but were opposed to any such uses of this Island," interestingly forgetting to mention the prospectus issued by Peter B. and Augustus Porter in 1825 which specifically invited manufacturing interests to develop the Island, suggesting the primitive woods was available to be cut down for fuel, the east end "might be covered with machinery," the west end "might be converted into delightful seats for the residence of private gentlemen, or appropriated to hotels and pleasure grounds" (Porter, in Adams, 1927). The rather self-serving characterization of the loss of Prospect Park and the Terrapin Tower to "other developers" has already been discussed above.

The Evershed plan — It is here where Peter A. Porter, Jr. revealed a power development scheme proposed by Thomas Evershed— the same engineer who produced the survey map of 1883 delineating the geographical details of what was to become the Niagara Reservation.

"In 1877 the idea of the great hydraulic tunnel had been matured by Thomas Evershed. His plan and proposition was to have the outlet of this tunnel at the base of the slope directly under Goat Island, extending the tunnel eastwards under the Island and then under the bed of the river; placing the mills on the main shore and connecting their wheelpits with the main tunnel by lateral tunnels" (Porter, 1900).

This is an interesting revelation because the Evershed tunnel, as excavated by the time of the essay's publication, was conceptually the same, only the mainland mills were at the east end of a tunnel, not running to the base of Goat Island, but at the base of the gorge at a point just north of the old Steel Arch Bridge by the American Falls, and somewhat south of the new Rainbow Bridge today.

Thomas Evershed

The year 1877 is a very interesting date, because it was relatively soon after this time that Lord Dufferin met with Governor Lucius Robinson and conveyed to him the idea of an international Niagara Falls Park. In 1879, Governor Robinson conveyed this idea to the New York State Legislature. In 1877 "the [hydraulic] canal and the property and rights belong thereto were purchased by Mr. Jacob F. Schoellkopf and Mr. A. Chesborough who organized the present company" (Van Cleve, 1903), that is, the Niagara Falls Hydraulic and Manufacturing Company—rival to what would become the Niagara Falls Power Plant. It was also near this date that the Porters found that they were in the process of losing their title to Goat Island, by having to lose it "owing to a partition suit now in progress" (Gardner, 1880). "The first move to protect public waterway rights in Ontario was ... taken in 1881. This was the passage by the government of Sir Oliver Mowat of the 'Rivers and Streams Bill.' The measure was the first step on the long road to public ownership of the Province's hydro-electric power resources. The Bill was disallowed three times by the government in Ottawa before the jurisdictional dispute was settled in Ontario's favour in 1884 by a decision of the Privy Council ... Following the Privy Council's decision, the Ontario Legislature in 1885 passed what became known as the Niagara Falls Park Act, a measure designed to preserve the natural scenery about the Falls" (Kiwanis Club, 1968).

Also in 1881, an electric generator to run arc lights was installed by the Brush Electric Light and Power Company at the Schoellkopf mill complex.

Peter A. Porter, in his history of Goat Island, indicated that the Evershed plan of 1877 was foiled by the preliminary legislation leading to the Gardner survey and report published in 1880. Yet Porter says that by not building the Evershed scheme on Goat Island, but rather locating it "to its present location," just upstream of the inlet canal for Schoellkopf's hydraulic canal at the east boundary of the Niagara Reservation, a change was effected "that resulted financially to the benefit of the gigantic enterprise." Without mentioning it specifically, this enterprise was the Niagara Falls Power Company, of which he was then an Officer and Director (Adams, 1927). Porter does not indicate the basis for the benefit derived from such change, but a guess might be made that greater efficiency could be designed into the configuration of the power plant in its subsequent location.

The Canadian connection — An important event which occurred in 1899 was the passage of a bill in the Ontario Legislature (62 Vict. 2, cap. III, par. 35) which gave the Commissioners of the Queen Victoria Niagara Falls Park in Ontario "authority to negotiate with the Canadian Niagara Power Company for the surrender of its exclusive right to use the water of the Niagara River in the Park" for the generation of electricity (Way, 1946). The Canadian Niagara Power Company was a fledging subsidiary of the Niagara Falls Power Company (Adams, 1927). In 1892, the Commissioners of the Canadian park had signed an agreement with the power company representatives for development of the subsidiary. Under the contract of 1892, the power company was given the exclusive right to develop power in the Canadian park for one hundred years "restraining any other company or person from using the water of the Niagara River, within the Park, for power purposes" (Way, 1946).

The problem was, probably for want of capital, the power company could not meet the requirements of the agreement to be able to produce power by the deadline of November, 1898, nor November of the next year by granted extension of the deadline. By 1898, the Commissioners had reason to wish to terminate the agreement altogether, as more lucrative agreements were in the offing with other capitalist interests hoping to develop their power companies within the Canadian park. Although the Canadian Commissioners could not terminate the agreement with the Niagara Falls Power Company, "requests for franchises to generate power had been received from other companies and it was generally felt that the monopoly clause should be canceled" (Way, 1946).

One consequence of the negotiations to extend the deadline for the generation of power for the Canadian subsidiary of the Niagara Falls Power Company, was the resignation of Casimir Gzowski, first chairman of the Queen Victoria Niagara Falls Park Commission. Mr. Gzowski, a civil engineer and Superintendent of the Public Works Department of Upper Canada, had been approached by Premier of Ontario Oliver Mowat in 1884 for the chairmanship of the proposed park commission and, in 1885, actually appointed him. Mr. Gzowski had directed the negotiations with power company interests from the start. "In 1892, deeply offended when Premier Mowat overruled him and allowed a lapsed application for a power franchise to be extended, Sir Casimir, pleading ill health, offered to resign as chairman" (Seibel, 1985). By 1893, his resignation was accepted. The year 1892 was notable for the contract made with the Niagara Falls Power Company granting its subsidiary, the Canadian Niagara Power Company a one hundred year monopoly on water diversion rights in the Queen Victoria Park.

With the powers granted by the Ontario Legislature, the Commissioners broke the monopoly of the Niagara Falls Power Company, and immediately gave the go-ahead to the Ontario Power Company which was waiting in the wings, but could not build a plant in the Canadian park due to the Commissioner's previous commitments to the Niagara Falls Power Company. Since the power company interests (Niagara Falls Power Company) were now on the American Commission of the Niagara Reservation, that body could lodge criticisms against the policies of their sister Commission on the Canadian side ostensibly in the interests of pursuing the mandate of the legislation to establish both parks: to preserve and restore the natural environment. It gave the power company a platform, under the guise of preserving the natural character of the cataracts, to criticize the policies of the Canadian Commission when these ran opposed to the company interests on the American side. The power company interests, in other words, were appointed to the Commission of the Niagara Reservation at a time when they had lost their monopoly to develop power in the Canadian park.

17th Annual Report of the Commissioners of the State Reservation at Niagara for the fiscal year from October 1, 1899, to September 30, 1900. Transmitted to the New York State Legislature March 13, **1901**.

In its next annual report (17 Ann Rep Comm, 1901, for 1900), the Commission announced the election of yet another power company executive to the board, the Peter A. Porter, Jr. discussed above, as treasurer and secretary. Mr. Porter's position was an elected one, not appointed, by a vote of the Commissioners, four of whom appeared to be working in concert, with Green a dissenting vote. For the first time, the secretary—treasurer has a voice, because Mr. Porter composed an extensive essay to accompany the reports of the Commissioners and the Superintendent.

Reference was made to the commitment to the projects undertaken in previous years and work "in accordance with a previously adopted plan," i.e. the Olmsted-Vaux plan.

Now, first reference is made to an adequate outlay of money from the Legislature, which "appreciating well the importance of providing for the great number of people who visit Niagara, made liberal appropriations for improvements intended for the accommodation and safety of the public." Not only were the appropriations liberal, but the Commission felt confident in asking for $20,000 over its "ordinary maintenance budget." Again, the Commissioners requested extra money to install the electrical equipment needed to avail themselves of the free electric power committed to the Reservation from the Niagara Falls Power Company, a total additional request of $47,265.15.

No reference is made to the possible degradation of the cataracts by the diversions of the power companies. Andrew H. Green, however, signed the report as President of the Commission.

In the report of the Treasurer and Secretary, Mr. Porter could yet refrain from his role as historian and further enlightens us on the history of the movement to free Niagara. He mentioned that the "agitation for the sovereign ownership at Niagara" began in 1877 (the year of the first Evershed plan, mentioned in his essay on Goat Island) but gives no reason for this date. One would presume the agitation began in 1879 when Lord Dufferin contacted Governor Robinson. Mr. Porter gives pride of place for the idea of preserving the Falls to two Scottish visitors, in 1834, and thereby bypasses the contemporary source of the attribution, whether Lord Dufferin, Governor Robinson, Frederick Church or Frederick Olmsted.

Porter wrote that the only access to the talus slopes between the American and Horseshoe Falls was the Biddle Stairs, "in honor of Nicholas Biddle of United States Bank fame, who is credited with having suggested to the owners of Goat Island this means of accomplishing the descent, and is further said to have offered to share the expense of their construction, although this philanthropic part of his suggestion was declined."

Again, as the Porter's frequently affect when writing history, they refer obliquely to themselves—in this case, "the owners of Goat Island." Nicholas Biddle (1786-1844) was apparently a friend of the family. Mr. Biddle, like Albert J. and Peter A. Porter, Jr., was a gentleman historian. "In 1820 and 1811, Nicholas Biddle, with the assistance of Clark and George Shannon, prepared from the original journals of Lewis, Clark, Ordway, Gass and others what remains the best account of the Missouri-Pacific expedition. Published in 1814, under the supervision of Paul Allen in Philadelphia, the Biddle edition has been many times reprinted, notably with elaborate critical apparatus from the original journals by E. Coues as History of the Lewis and Clark Expedition (3 vols. and atlas, New York, 1904-5)," (Brebner, 1955). The Biddle Stairs was a wooden tower with spiral stairs inside. It was built in 1818 (Adamson, 1985). Out of one of its windows half way to the bottom, Jacques Gerard Milbert, who was "sent to America by the French government to study the natural history of the eastern states" between 1815 and 1822, painted a view of the falls (fig. 19 in Adamson, 1985).

Mr. Biddle, a financier, was appointed a government director of the Bank of the United States by President Monroe in 1819, and became conspicuous for the attacks on him by Jacksonians, in their attempts to destroy the Bank (Bridgwater & Kurtz, 1963).

The Porters perhaps never lost their association with the Biddle family, for a Nicholas Biddle served in 1918 as a director of the Niagara Falls Power Company (Adams, 1927).

Railroads, trolley lines — Peter A. Porter, Jr. reported that, in addition to the granting of the license of the previous year to the Niagara Falls and Suspension Bridge Railway and associated rail lines, the Commissioners had granted a license to the International Traction Company "(which controls largely the trolley lines on both sides of the river at Niagara, and also two bridges across the gorge) to lay a track on the Riverway between Falls street and the north end of the reservation." Green's initiative in putting a bridge across the Niagara gorge below the whirlpool for the benefit of tourist interaction with the natural attractions of the area had been superseded by the interests centered in the Niagara Falls Power Company. This company had dominated local transportation routes on both sides of the river, dominated the shoreline of the upper river for their industrial purposes, received numerous concessions to develop water power without compensation to the Commissioners of the

Niagara Reservation, and now, since 1899, controlled that Commission and began to use it to further their aggrandizement of economic power at Niagara Falls. Their influence on the Canadian side of the river should not be overlooked. The province had been receiving benefits derived from the company's commercial infrastructure in terms of transportation (trolleys, bridges) and power development.

Peter Porter recommended that "a topographical survey of the Falls be authorized and provided for. The value of the survey made ten years ago, and furnishing in connection with several prior surveyors a minute and accurate record ... of the changes, erosive and creative, in the lines of the banks, renders this recurring ten-year period ... a suitable time for this important work."

Ten years previously, Andrew Green had written to the Hon. John Bogart, State Engineer and Surveyor, stating that Bogart had sub-mitted to Green a letter of introduction, an assistant of his requesting "in-formation and directions concerning the surveying work at Niagara" be-ing undertaken under Chapter 570, Laws of 1889 (Green, in 7 Ann Rep Comm, 1891, for 1890). Green was anxious for the survey to begin, for it would assist him in prosecuting for a bridge to be constructed to Canada. Green further asked Bogart and his assistant questions on the condition of erosion on the south side of Goat Island, and information on the char-acteristics of the "tunnel about to be constructed", that is, the hydraulic tunnel of the Niagara Falls Power Company. What Green apparently did not know was that Mr. John Bogart was "retained as consulting engineer to prepare a detailed and large-sized map of the location under considera-tion" for the development of the generating plant of the power company. Mr. Bogart, from 1890 to 1893, served as secretary and consulting engi-neer to the Cataract Construction Company "Niagara Advisory Board of Engineers," which was the holding company for the Niagara Falls Power Company (Adams 1927). Mr. Bogart, working for both the State and the power company, much like Evershed before him, prepared, for the latter, "accurate surveys of the land with the view of possible purchase, and of the lands under water with riparian rights" (Adams, 1927). Mr. Bogart calculated the volume of flow in the Niagara River in 1890 (7 Ann Rep Comm, 1891), and in 1900 he again made measurements of the volume of flow (Adams, 1927). Mr. Bogart was a stockholder in The Cataract Construction Company, which managed the Niagara Falls Power Com-pany.

In the seventeenth annual report, 1901, for 1900, the new rail-way on the Riverway is further considered. To indicate the value of this rail extension into the Reservation, one might consider the revenues anticipated from the upcoming Pan-American Exposition by a quote from the nineteenth annual report (1903, for 1902): "the amount of money brought into the State and left here by tourists drawn hither from other States and countries is very large; and merchants, hotels, railroad companies, and many other interests derive a large revenue from the patronage attracted by the Falls. A single railroad company, for example (the New York Central and Hudson River), received $170,000 in fares during three summer months of 1902 on account of its Niagara business alone." 1902 was the year subsequent to the Pan-American Exposition.

The electric railway authorized by the Commissioners over Green's objections "between Falls avenue and the entrance to the International Steel Arch Bridge", which bridge was owned by the company granted the license to operate the cars, caused some problems. Due to the presence of the electric rail, "there is not the entire freedom for vehicles formerly enjoyed in that part of the riverway, but aside from that the Reservation has suffered no injury. The operation of the cars as a rule, has been satisfactory, proving the wisdom of the State in retaining the ownership of the track and the absolute control of the operation of the cars within the distance traveled inside the Reservation" (Welch, in 17 ARC, 1901). This "wisdom" apparently derived from power company interests managing the Reservation through the Commission to promote their transportation interests, thanks to a special law permitting construction of railway facilities in a public park, as discussed above. A structure, "dignified and substantial with architectural character worthy of so prominent a location", that is, on the riverway, is proposed to be built for and by the Electric Railway Company.

Buildings — Vaux and Emory's building, built in 1896 at the base of the inclined railway, "has been repaired in places where it was damaged by the ice during the winter. April 16th the gable on the northern side of the lower terminal station was torn off by the ice. The damage has been repaired" (Welch, in 17 Ann Rep Comm, 1901, for 1900). These repairs appear to have been anticipated by Welch in the year the building was built (Welch, in 13 Ann Rep Comm, 1897). In 1899, "the roof of the lower terminal station of the Inclined Railway which had been damaged by ice" was repaired (Welch, in 16 Ann Rep Comm, 1900, for 1899).

A new building was proposed to replace the one earlier burned by an arsonist to house the office of the Commissioners and the Superintendent. Here "should be located all the offices of the Commissioners and officials, and which would also furnish a much needed Shelter building for the accommodation of visitors, which building should be located in Prospect Park" (Porter, in 17 Ann Rep Comm, 1901). The building would be the design of the State Architect, Mr. G. L. Heins. Mr. Porter wrote that originally the design was a "beautiful Gothic structure" but subsequently "certain changes were made reducing the height but not the ground floor area of the building and still preserving its classic beauties" (Porter, in 17 Ann Rep Comm, 1901).

This claim to the "Gothic" character of the administration building by Mr. Porter is very curious for one as cultured as he was. The actual building is of a strikingly neoclassic style, complete with Greek columns, pediment, frieze and classic Greek roof ornamentation, such as the acroterion on the front of the building. There is nothing remotely Gothic about it. The building was built to face the incoming visitors on the trolley line depot directly in front of it on Riverway Street. It was one stop on the line which included the transformer houses on the track from Buffalo, bringing excursionists to Niagara from the Pan-American Exposition.

It is almost as though Porter were deliberately introducing a double meaning, that the original Gothic structure, which would have represented the picturesque character of Victorian architecture, had been modified without loss of its "classic" beauty, that is, its Greek elements.

Since the building was built to accommodate the Pan-American Exposition, it might be instructive to compare this administration building with the "New York State Building" designed by Buffalo architect George Cary for the Pan-American. Its "American Renaissance or Neoclassical Revival style of architecture" is parallel to that of the Reservation's administration building, complete with acroterion. Earlier, the neoclassical style had been the dominant architectural theme of the Columbian Exposition buildings in Chicago in 1893, on which McKim, Mead and White worked (Newton, 1971), and was carried over into the Pan-American (Brown & Watson, 1982). Curiously, the architectural theme of the Exposition was not classical, but the baroque of the Spanish Renaissance and the New York State Building, neoclassic in style, was the only building intended to remain after the Exposition was over (Brown & Watson, 1982),.

The New York State Building is now the home of the Buffalo Historical Society, having been dedicated in 1902, the year after the Exposition.

Earlier, in the 15th annual report, it was Welch who promoted the "urgent necessity of a substantial administration building" in Prospect Park, which would include "waiting rooms, lavatories, toilet rooms, store rooms, a parcel room, and other conveniences for the accommodation of a large number of people."

Apparently, an invitation had been extended to perhaps more than one architect or architectural firm to submit designs for this building. It is probable that Green, as President of the Commission, had extended this invitation. The probability arises due to the reaction by other members of that body, Welch and Alexander Porter, which resembles the Commission's earlier attitude toward permitting a private railroad company to build lines, a terminal and depot within the Reservation on Riverway. In that circumstance, Green appeared to be the only dissenting vote, as discussed above.

In the Daily Cataract, however, a newspaper of the City of Niagara Falls, it was reported in October 26, 1899 (p.1) that "Heins [the State Architect] was asked to come to Niagara at the special request of A. J. Porter and Thomas Welch to confer on plans" for this building.

On November 1, 1899 (p.1) in a message to Thomas Welch, "Mr. Heins stated that the law was very explicit in proscribing his duties [as State Architect] and that it is absolutely necessary for him to prepare the plans for the new State building."

"Mr. Heins did not discuss the question of the loss of the architects who submitted plans for the building any further than to state his reasons for sending them back to the Commissioners." On December 7, 1899 (p.1) a meeting was reported of the Commissioners with Heins regarding the shelter and administration building. Five plans for the shelter building were to be examined by the State Architect, but "he considered it improper for him to examine the plans and returned them unopened, saying the law required him to make the plans." Superintendent Welch had "a copy of the law relating to the duties" of the office of the State Architect. The final design of the present building is attributed to Heins.

The Commission was anticipating "large crowds which are expected to visit the Pan-American Exposition, and therefore the Reservation in 1901" (17 Ann Rep Comm, 1901, for 1900). The new shelter building was completed July 4, 1901: "it affords many conveniences and accommodations for the public, as well as offices for the administration of the affairs of the Reservation" (18 Ann Rep Comm, 1902). The building contained toilets and storerooms and was constructed on the Riverway, in Prospect Park (Welch, 18 Ann Rep Comm, 1902) and probably was meant to be used in conjunction with the trolley depot, which it faced.

The design for the Shelter Building in Prospect Park marked a radical departure from the rustic designs of Downing Vaux and Emory.

Although designed by the State Architect, G. L. Heins, it was built in the style of Stanford White (Otis, 1982)—Peter A. Porter, Jr., as noted above, describing it as "Gothic."

Stanford White was a partner in "America's most prestigious," "America's best known architectural firm, McKim, Mead and White" (Fox, 1986). He designed the Williams-Pratt House (1895-1896) on Delaware Avenue in Buffalo, New York, in a Georgian Revival style, and the William-Butler house "next door." The former house is on the National Register of Historic Places. In Niagara County, in Lockport, New York, also on the National Register, is Union Station, 1888, depot for the New York Central Railroad, built in a Romanesque-Revival style. The architect is actually unknown but believed to be Mr. White.

The third National Register building designed by the firm of Stanford White is the Adams Power Plant Transformer House, 1895, in Niagara Falls, New York (Fox, 1986). This is one of the power houses of the Niagara Falls Power Company. "The massive canal power house is a handsome building, designed by Stanford White, and likely to stand until Niagara, spendthrift fashion, has consumed its way backward, through its own crumbling strata of shale and limestone, to the base of it. This building is outwardly of hard limestone and inwardly of enamel brick and ordinary brick coated with white enamel paint," (Martin 1896, quoted by Dow, 1921). Indeed, Power House No. One looks strikingly like the photograph of the Boston Public Library (ca. 1908) designed by White's partner, Charles F. McKim, which was said to relate to a theme of the Italian Renaissance (Newton, 1971). The works of McKim, Mead and White have been characterized as marking the beginning of "mercantile classicism" in industrial design (Newton, 1971). The staid designs of this firm, influenced by the brilliant Romanesque architecture of Henry Hobson Richardson, terminated seventy-five years of picturesque romanticism, often "carried to ridiculous extremes" of eclectic borrowings of architectural elements from different eras and different buildings in a single dwelling (Newton, 1971). The turning away from Downing Vaux's tradition of work at the Reservation to that of McKim, Mead and White reflects this change in North American architecture from romanticism and perhaps sentimentality, toward the modern, commercial and practical. So far as the present study has been able to reveal, no architect associated with Central Park was ever consulted again for the Niagara Reservation although the Canadian Commission continued to consult with Frederick Law Olmsted, Jr., as did the federal government later around 1906.

More characteristic examples of the rustic and picturesque in park architecture may be seen in those constructed in the Queen Victoria Niagara Falls Park in the first two decades of its existence. The change from picturesque to pragmatic in Reservation buildings after 1898 also reflects the loss of Green's control of the Commission of the Niagara Reservation, and its ties to the architecture and landscape influence of Central Park, to Olmsted, Parsons and Green himself. The primitive wildness of the Reservation and the Queen Victoria Park, which was to have been preserved in law, was in harmony with picturesque and rustic architecture. Change to the strongly architectonic and solid architecture adopted by commercial interests, such as the power companies at the cataracts, reflects the new Commission's interest in conveying to the public the impression that the falls was ancillary to the power companies, only one of their souvenir attractions, like its power houses and its Shredded Wheat factory and its modern electric trolley system (see below).

Powerhouses — The White power house was not viewed with respect by all visitors. It was roundly bashed by H. G. Wells in 1906, whose opinions partake of the astringent quality of Lord Kelvin's singularly ugly valuation of the cataracts, quoted by Green in the 14th annual report for the year 1897 (1898), as noted above. Wells wrote:

"I turned back to look at the powerhouse as I walked towards the falls, and halted and stared. Its architecture brought me out of my daydream to the quality of contemporary things again. You know, it is such an inconceivably dull piece of building—a box of bricks exterior for these engineering splendors—a shock, a scandal like a bowler-hat on the king of kings. What an architect! I'd almost as soon have had one of the Schoellkopf sheds," referring to the squalor of the Canal Basin on the cliffs below the falls, even then coming under the scrutiny of the Secretary of War, Howard Taft (Wells, 1906).

As the design relationship of the new Commissioners of the Niagara Reservation with Mr. White may take on more significance later in this narrative, perhaps it will be instructive to provide the commercial rational for the appearance of this power house designed by McKim, Mead and White of New York.

According to Adams (1927), the power houses to be erected soon after the establishment of the Reservation, and just upriver to its eastern boundary, were, in their design, to transmit certain characteristics to the viewer. These characteristics were specified to the architects when they were approached to design this power-house.

The power-house must be attractive and embody the essence of "grandeur," it was to be dignified, impressive, monumental." They were essentially to rival the glories of the cataracts such that "souvenir pictures carried away by the visitors should include one or all three of the power-houses" (Adams 1927).

The houses should be "protective," reminiscent of severe winter seasons threatening to shut down, by ice, the hydroelectric operations on which the customers of the utility were to depend. The structure should inspire confidence and "afford shelter."

"The power-house should be instructive and afford a demonstration and an exhibition" to the visitor, the tourist, who would wish to observe the great machinery which delivered power from the river over great distances to motivate the trolley cars, electric lights and other appliances the public may come to take for granted.

It should "express in its design the purpose of its construction." As part of a massive public education campaign, the houses would have provision made "for the admission of the public without danger to themselves and without interruption to the service, and for intelligent and helpful guides who would explain the methods of utilization and distribute the literature provided. Each visitor should learn enough to desire to know more, and should be impressed by the evidence that the managers of the enterprise had confidence in the investment of their own capital, and had built for endurance and continuity of operation."

As will be discussed below, power company interests were attempting to transform the image of Niagara Falls as a magnificent feat of natural processes deserving protection by the governments of New York and Ontario from encroachments by commercial interests, into a glorious appendage to something even more triumphant: technology. As if in imitation of the campaign to establish the Reservation during the first half of the 1880's, the power company was hiring guides and distributing leaflets explaining, and also glorifying the technological advances that made possible the transformation of industrial power from steam to electricity, using the diverted waters which once had cascaded over the brinks at the cataracts.

The Porter family had put its clear stamp on the company's architectural design: "by the use of native stone, as the oldest inhabitant had recommended by building his own home of that material, and keeping the surface of the stone roughly trimmed, the character of the structure would be defensive, as against the storms without, and protective of the valuable machinery enclosed therein" (Adams, 1927). Again, the peculiar oblique reference to the "oldest inhabitant;" such references in the literature generated by Albert and Peter A. Porter always refer to their own family. The oldest inhabitants were apparently not meant to be the Indians, and probably not John Stedman. That left Augustus Porter who lived near the Stedman estate on the American mainland across the river from Goat Island. If the reference is truly to the farm building of Augustus Porter, then all related structures built according to this design were a monument to the Porter family and to the success of the elaborate power and transportation infrastructure girdling the cataracts on both sides of the river toward which they had contributed so much.

As a matter of fact, there is something left today of the Stedman buildings, reflecting the structures made during the French and British occupation of lands at the cataracts—a stone chimney standing in a shaded location beside the Carborundum Building on Buffalo Avenue beside the Robert Moses Expressway. Welch referred to this chimney as still standing when the Reservation was established, indicating that it was the "chimney of the Stedman House, built in 1761 (2 Ann Rep Comm, 1886). There is a metal plaque on it today which reads:

"Built by French, 1750, 100 ft. westward in Fort Little Niagara's barracks, which they burned in 1759. To it, British built in 1761, the Stedman House (where that master of the portage lived until U.S. occupation in 1796), which in 1808 became Broughton's Tavern. Burned by British in devastation of 1813. Re-erected here in 1898 by Niagara Falls Power Company. Marked by the Niagara Frontier Historical Society in 1915 [sic]."

The relevance of this architectural character and the impression it was to make on the visiting populace to the Niagara Reservation is that many buildings erected on the Reservation today are of rough-hewn stone of the character described for the power company power-houses, in unwitting monumentalization of the Niagara Falls Power Company if not the ancestral home of Judge Augustus Porter.

"The public carriage stands on the reservation were abolished May 26th in accordance with the resolution adopted by the Board. A copy of the resolution was mailed to each driver of a licensed carriage. The drivers have complied with the requirements almost without exception" (Welch, 17 Ann Rep Comm, 1901, for 1900). The reason for this appears to have been due to the "licensing of improper persons by the city government." Since the carriage stands were reinstalled after the Pan-American and had not been abolished before or since, one wonders if the stands were abolished so as to ensure a monopoly on both use of the trolleys and the Reservation carriage service for this brief period when hundreds of thousands of dollars would be made by transportation companies serving the momentarily expanded population. Welch would criticize the trolley line for packing its cars with more people than a car could conveniently handle (18 Ann Rep Comm, 1902).

As if to further increase the visitor's dependence on the trolleys, the Reservation carriage service was instructed "to deliver passengers only on the river side of the Riverway and to use the Reservation carriages for hire only within the Reservation. The Reservation carriage service should be operated without combination of any kind with any other local business enterprise" (Welch, 17 Ann Rep Comm, 1901).

There were no disbursements to A. H. Green for travelling expenses in 1900, nor for the next year, 1901, nor for 1902.

18th Annual Report of the Commissioners of the State Reservation at Niagara for the fiscal year from October 1, 1900, to September 30, 1901. Transmitted to the New York State Legislature February 24, **1902**.

The next year, that of the Pan-American Exposition of 1901 (18 Ann Rep Comm, 1902), the new Shelter building had been finished in Prospect Park, with its accommodations for the public, and administrative offices. Three million visitors were calculated to have visited the Reservation in that year.

The station or depot for the Niagara Falls and Suspension Bridge Railway called by Welch the Electric Railway Company, which, as I have discussed elsewhere, was probably a shortened name for the Buffalo and Niagara Falls Electric Railway Company, on the Riverway in the Reservation, had been completed and "A double track has been laid in the riverway from Niagara street to the entrance to the steel arch bridge, and a second track in the riverway from the entrance to the steel arch bridge to the electric railway station, the street having been widened for that purpose, according to the agreement with the Electric Railway Company" (Welch, in 18 Ann Rep Comm, 1901).

Welch complained that the sidewalks were blocked by the trolley cars, that the trolleys were dangerously fast. Welch differentiated between a station and a railway terminus, that is "the usage of the riverway by the company as a terminus and standing place for cars, in violation of the agreement."

Visitors, in spite of the station being located in the Reservation, were unloaded "at the corner of Falls and Prospect Streets, the headquarters of many confidence men and gamblers, where the lack of order and protection to visitors gave rise to many serious causes for complaint," and Welch warned that if the company continued this practice, their license would be revoked. The license was never revoked, indeed, in the very next year the civil government of Niagara Falls went out of their way to correct the ephemeral evils established adjacent to the Reservation to take advantage of the sudden efflorescence of visiting peoples. "During the past year [1902] a faithful discharge of duty ... by the Mayor, the Common Council, the Police Commissioners and the members of the police force, has accomplished results so gratifying and creditable to the city, that a relapse into the condition of affairs which existed in 1901 is extremely improbable" (Welch, in 19 Ann Rep Comm, 1903, for 1902). The crowds with their money were gone, so was then the free-for-all fleecing of the public apparently sanctioned by local government and decried by Welch in the previous report.

No public carriage stands had been permitted on the Reservation during the year of the Pan-American, leaving visitors the option of either using the trolleys or the carriage concession granted by Welch on Reservation grounds. The carriage stands along the Riverway by the trolley station were reinstated only after the millions had departed, having had to leave the Reservation grounds to find a conveyance not controlled by the trolley interests or that of the Reservation carriage service.

The contract between the Niagara Falls and Suspension Bridge Company and the Commissioners of the Niagara Reservation for the hasty construction of a second track in the Riverway is appended to the end of this report. The original contract of 1899 was for a single street railway track. With the building of the trolley station, additional construction was requested by the company for "turnouts to its passenger station fronting upon said Riverway" and illustrated on a map accompanying the report.

The contract was signed by the Commissioners of the Niagara Reservation George Raines, T. P. Kingsford, Charles M. Dow and Alexander J. Porter. Oddly, the former Secretary of the Commission, Richard F. Rankine, signed his name to this document as secretary-treasurer, but this time for the International Traction Company, a post for which he had resigned his office on the Commission (see discussion above).

Andrew Haswell Green, President of the Commission, did not sign his name to this contract either.

Illumination — In Canada, "During the Pan-American Exposition which was held in Buffalo in 1901, the Falls were lighted with searchlights in an attempt to attract attention away from the Exposition at night. One of the searchlights used in this illumination was mounted beside the Michigan Central Railway station at Falls View. Power for this searchlight came from a wire connected with the International Railway Company electric street car line located directly below. It is only a coincidence that their Royal Highnesses the Duke and Duchess of Cornwall and York, later King George V and Queen Mary, visited Niagara Falls during the time of this display" (Seibel, 1985). Given the powerful financial and technological connections with Great Britain and Europe by power company interests of the time, as will be discussed elsewhere, this event may in fact not have been entirely unplanned. The Falls View Station was just above the International Railway Company Powerhouse, formerly the powerhouse of the Niagara Falls Park and River Railway Company (see Pan-American Exposition map and 1896 historic map in Seibel, 1985 opposite p. 34). This railway, and the International Traction Company, were managed by Mr. Caryl Ely.

Some of the illuminated attractions of Niagara Falls were not in competition with the Pan-American, but in concert with the Exposition. "The illumination of the Pan-American Exposition was designed with the Electric Tower as the focal point. The tower was used as a searchlight, signaling at night. The signals were answered by an observation tower located miles away in Niagara Falls" (Brown & Watson, 1982). This observation tower was probably the Moose Tower Hotel, located on Riverway beside the trolley terminus. Dismantled in 1904, it went to the Louisiana Purchase Exposition in St. Louis "where it served as a wireless telegraph station."

19th Annual Report of the Commissioners of the State Reservation at Niagara for the fiscal year from October 1, 1901, to September 30, 1902. Transmitted to the New York State Legislature February 18, **1903**.

The nineteenth annual report of the Niagara Commissioners (1903, for 1902) is an anniversary edition celebrating the twenty-year anniversary of the Commission of the State Reservation at Niagara. It shows, in the activities authorized by the Commission, *an increasing toleration of works of development*. For example, the "cutting away and grading of the [natural] banks at the approach to Goat Island." All approaches to features, such as the Three Sisters Islands, throughout the Reservation were to receive this treatment in the decades ahead, leading to an accumulation of topographic modifications such that the original topography of much

of Goat and other islands would become altered. A pride in the technological improvements offered by the day characterized the administration
over its commitments to restoring the primitive environment of the Reservation, along the pattern of that still in existence on Goat Island in the
1880's and '90's.

Improvements in the technologies of transportation, of the furnishing and distribution of water, sewage and electricity—all concerns of
the power company, are applied to Goat Island. Environmental concerns
take less priority. The Commission was in the business of "public convenience" as, indeed, was the electric utility corporation, which was, by
its very definition, in the business of being useful to society, as were and
are gas, water, telephone and transportation utilities.

In 1901, the Cottage dwelling house was removed from the entrance to Goat Island—this was probably where maintenance facilities
were concentrated, such as the nursery and lumber yards, and where
workmen were headquartered. The Commissioner's Office building was
finally removed from Green Island. Grading continued to occupy the
administration—destroying the natural character of Prospect Point,
Stedman's Bluff and Terrapin Point.

The State Engineer and Surveyor had recommended a system of
electric lighting for installation in the Reservation, and the Commissioners requested $10,000 for this.

Under the new Commission, the cost of running the Reservation
began to spiral upward. There was an abandonment of the assurances,
made when the Commission was run by park interests, that a few structural placements of superior construction, and a return of the environment to, essentially, the maintenance by natural processes, would present
a park with little annual fiscal requirements. Indeed the revenue from
such facilities as the Cave-of-the-Winds and the conveyance to the base
of the cliff at Prospect Point, then the Inclined Railway facility in the
1900's, would have made operating the Reservation require little additional funds.

The new Commission, however, requested $10,000 more for
maintenance above the $25,000 they had asked in previous years. "The
amount will be more inadequate in the future, because the care of the
new shelter building requires additional labor. The new stone arch bridges to the islands are not provided with gates, thus requiring the service of
watchmen night and day, and the increasing number of trees and shrubs
under cultivation requires an additional amount of care and maintenance.
The proper maintenance of the Reservation roads and walks requires a
large annual expenditure."

A historical review of the Reservation — Again, the Legislature is treated to a recapitulation of the "Creation and Development of the State Reservation at Niagara" on the twenty year anniversary of the Commission (not the Reservation). The anniversary was in 1903, commemorating, not the year 1885 which had received such fanfare with Governor Hill's passage at the very last minute of the act to appropriate the money to pay for the legislation Governor Cleveland had enacted in 1883: to pay for the Reservation.

The legal and political requirements to continue the mandate given to the Reservation Commission which evolved between 1879 and 1885 also happened to be almost perfectly matched with the legal and political requirements of the power companies. The preservation of Niagara Falls was a fine regulatory mechanism which could be used politically to block or promote rival power-development interests, whether upstream in the Great Lakes, across the river in Canada, or adjacent to the company, up or down river. This potential of the Reservation was very well known by some, which accounts for the "packing" of the Commission with power company interests under the latter part of the Green presidency. The chief difference between park interests and power company interests was one mile: either keeping the river volume intact until it fell over the brink, or keeping it intact until the water could reach the intakes and be diverted a mile or so upriver from the cataracts.

The power companies, the Niagara Falls Power Company and its rival, the Niagara Falls Hydraulic and Manufacturing Company, were in a position where they had to compete with waters in the Niagara River with American interests in the upper lakes, such as the construction of the Chicago drainage canal which has posed a threat to power generation at Niagara up through the twentieth century (Moses, 1960), a canal proposal "across the State of Wisconsin from Lake Michigan to the Mississippi, and another for a canal from Lake Superior at Duluth to the Mississippi at Minneapolis," and Canadian enterprises such as the Welland canal "and the proposed canal connecting the Georgian Bay with the Ottawa River." One must not forget the federal intent to build a dam at Buffalo to raise the water levels of the Great Lakes (16 Ann Rep Comm, 1900). American sectional transportation issues appear to have been at play here between Mississippi River states and the shipment of goods to the port of New Orleans, and Great Lakes States which sent their goods to New York and other cities on the Atlantic coast, in addition to the autonomous British-Canadian shipping routes.

Consequently, it was important for both park and power company interests to periodically review the history of the movement to preserve the falls, reaffirming the laws established to protect it and the associated objectives for the New York State Legislature. The interests of the companies could be hidden behind the "scenery," that is, the conceptual framework of Frederick Law Olmsted and his colleagues in the social reform factions of government. These concepts, so important in American social policy today, could be used rhetorically, a scenery behind which the private interests of the power companies could prosecute their agenda.

A national context — The 1902 retelling of the movement to preserve Niagara is placed within a national context. The principle of "governmental authority to protect and preserve extraordinary phenomena in natural scenery" was said to have begun "so far as we have been able to trace the genesis in this country" with Yellowstone National Park in 1872. This point of view completely ignores the situation at Yosemite, which Green in a previous annual report indicated had suffered from mismanagement (7 Ann Rep Comm, 1891). In that report Green noted that "so strong is the feeling that the Yosemite valley is a national treasure that it is not unlikely that the National Government, at no distant day, will resume the title to the property. A bill providing for the enlargement of the public area surrounding the valley was introduced into the Fifty-first Congress and passed during the first session."

It was very clear to many in the country that Yosemite was the first momentous act of government, both federal and state to established a park or reservation to protect "extraordinary phenomena in natural scenery." It is difficult to understand this oversight in scholarship.

It is because of this omission of fact that one might wonder what contribution Mr. Green had actually made to the writing of this essay. Green was, after all, a colleague of Olmsted, deeply involved in public parks and other civic institutions for the public welfare. The reader must assume Green was oblivious of Olmsted's role as effective head of the Commission to protect Yosemite. We must also assume Green was unaware of what he had written in his previous annual report where he outlined a public land history in which Yosemite was clearly first, followed by Yellowstone, and Niagara as the third great example of government preservation (7 Ann Rep Comm, 1891). The objectives of these three government acts was "to protect and preserve extraordinary phenomena in natural scenery from injury and desecration" (7 Ann Rep Comm, 1891).

One might use this peculiar omission to consider that Green did not write this portion, or any of the celebratory history of the Niagara Reservation in this annual report. Whoever wrote this history also used the third person pronoun "we" in reference to the Commissioners of the Niagara Reservation. No report written prior to the sixteenth report ever referred to the Commissioners as "we." Green always conveyed the ideas accepted by the Commission impersonally, as though he were a disinterested reporter. In the history here discussed, "we" or "our" is frequently written with respect to the activities of the Commission at the beginning of this essay. "The following is our estimate of the amounts required ..." (p.12), "in each year we have been obliged to dispense with necessary labor ..." (p. 13). When the author wrote, on page 14: "So far as we have been able to trace the genesis in this country of the principle of governmental authority to protect and preserve extraordinary phenomena in natural scenery, it was first established in 1872 ...," this statement might be believable as long as Green was not able to contribute his knowledge, as noted above, that is, if Green was not included in "we," in the present deliberations of the Commissioners.

Little or no reference is made to Green's abandonment of the Ontario and New York Legislatures, and the active work he had been pursuing at the Federal level: the American Congress. This is very peculiar, because Green had just triumphed in this regard. Buried away at the back of this report, which is supposed to celebrate and illuminate the activities of the Commissioners of the Niagara Reservation for two decades, the culmination of their achievements and bitter problems, were the "Documents Relating to the International Waterways Commission."

As will be seen, the Waterways Commission was to trigger a great contest in many arenas between State and Federal rights, park interests and power interests, public and private rights, all during the administration of Theodore Roosevelt, a reform-minded Republican, with Howard Taft as Secretary of War. Taft's experience in preserving the cataracts at Niagara would color his interest in forming a branch of the federal government to preserve great examples of the North American landscape, which would ultimately lead to the establishment of the National Park Service in 1916.

Did Peter Porter write it? — One other person on the Commission, its Secretary and Treasurer, Mr. Peter A. Porter, Jr., was demonstrably competent to write such an history. He was a professional historian regarding matters in Niagara Falls, New York, and had had printed several histories published by the Niagara Reservation Commission before he became an officer of it. His histories were published by the Commission due to Green's acknowledged interest in original research into all aspects of science and history which touched on Niagara Falls (Spencer, 1907). In Dean Adams' history of the Niagara Power Company (1927), Mr. Porter is identified in the index as an historian (although erroneously as the son of Gen. Peter B. Porter).

From the point of style, the history of the Reservation and the public movement to establish it presented in this Annual Report is devoid of the high moral and exalted tone frequently used by Green, and there is a muted treatment of the threat to the character of the Falls through diversions by the power companies in New York State, and of castigation of commercial interests which had characterized Green's previous annual reports. Emphasis is made on the conspicuous developments of power companies on the Canadian side, and the Canadian Commission's permission to have these companies built within park land. Photographs are included showing disruption of the lands the Ontario park was supposed to have been committed to preserve and restore. Additionally, there was the tragedy in which two workmen in the American Reservation were killed by debris blasted across the river from excavation for the Ontario Power Company being built in full view of the Niagara Reservation at the base of the Horseshoe Falls. Responsibility and concern are diverted from the Niagara Falls Power Company and the Niagara Falls Hydraulic and Manufacturing Company, and the expansion of their operations, to companies whose relationships with these two companies is obscured, to issues of national and international concern.

It must be noted that both Canadian power plants were controlled by American power companies, and these were rivals with one another—the one a subsidiary of the Niagara Falls Power Company, two of whose directors were officers of the Commission of the Niagara Reservation, and the other a plant built by the Niagara, Lockport and Ontario Power Company, financed in large part by capital in Buffalo (Brown and Watson, 1982).

Another curious fact is that a special edition of this Annual Report had been issued separately from the "ordinary" copies printed by the Argus Company of Albany for the New York State Legislature. Previously, the annual reports had been printed several times on the least expensive pulp paper. At any rate, the quality of the paper varied and was never exceptional. The special edition was exquisitely printed, by the same Argus Company, on what may be rag paper, or at least its rag content was the highest of any previous or subsequent publication of the Annual Report. It was printed with a special ink and the borders increased from 7/8 inch borders to something like two inches on a side. The paper had an elaborate deckled edge, to suggest hand-made paper.

What event could have spurred the Commission to issue such an elaborate book-form of its annual report, or to celebrate the Commission's existence, rather than the existence of the Reservation itself, in 1885? In 1900, it was announced that the name of one of the islands in the American channel was to be changed to honor the President of the Commission, then still active on the Commission and certainly still alive. This Commission also decided to give Thomas Welch, the Superintendent, a "superb silver loving cup" "in appreciative recognition of his faithful and efficient services," and he, too, was not retired and still alive. It is a tragic irony that both men would lose their lives later in the same year this special edition was issued, Welch by typhoid, Green by murder.

The motive of this commemoration was most likely to get the same members of the Commission re-appointed and reconfirmed in that year, when their five year terms expired.

Green, in earlier annual reports did not mark the time of the Commission, but of the Reservation itself—from 1885. For example, in the twelfth annual report (1896) he wrote: "The tenth year of the existence of the Reservation ended July 15, 1895. The changes that have taken place during this decade of years have been indicated in the annual reports." Again, in the fourteenth annual report, "the Reservation has been public property for more than 12 years (July 15, 1885 to September 30, 1897)." Green celebrated and defended the Reservation, not the Commission.

As to Yosemite, perhaps one reason the fact of Yosemite was "overlooked" was because the Federal Government had set a dangerous precedent there—they had found Yosemite, under the control of the government of California, to be degraded by commercial interests over which the State exercised little control, and the federal government had nationalized it. The boundaries, as Green mentioned in 1891, had been enlarged. If the boundaries of the Niagara Reservation were enlarged for the preservation of the cataracts, the manufacturing and power interests would be directly compromised. As a matter of fact, within the decade, in the midst of a political fight over diversion of Niagara's waters, Howard Taft, as Secretary of War, would in fact support a recommendation for the establishment of a national park at Niagara Falls (Taft, 1909).

The Federal Government had to be made preoccupied with only some of the concerns involving the degradation of the cataracts by commercial issues. By time Green lost control of the Commission with the new appointments of 1898, he had been inexorably led towards appealing to the Federal Government to regulate the proliferation of power concerns along the upper Niagara River as recommended to him earlier by the subcommittee to the Committee of Legislative Powers, with the hope of getting more power in the domain of the Commission, who at that time were solely interested in the environment at Niagara, to regulate diversion issues. That the federal government might see fit to nationalize the park at Niagara Falls, as they did at Yosemite on October 1, 1890, was a distinct possibility, and that the government might see fit to nationalize the great water-power generating potential for the benefit of its citizenry, especially since the role of another sovereign power was a crucial element in both preservation and power, was also a distinct possibility.

The road Green was taking was clearly disadvantageous to the Niagara Falls Power Company and other power interests at the brinks of the cataracts. But after the appointment of Alexander Porter to the Commission of the Niagara Reservation, on March 25, 1898, a Director and Trustee of the Niagara Falls Power Company, followed the year after through the election by the Commission of his relation, Peter B. Porter, another company Director, none of Green's arguments was heard of again—except with apparent alterations, as will be discussed below. Again, in the 1902 history presented to us, most probably by P. A. Porter, Jr., of the movement to preserve the cataract environment, we are treated to an account of the interests of the Porter family and their descendants in purchasing the area embracing the cataracts. The Porters "are entitled to the appreciative remembrance of the lovers of Niagara's beauty for their guardianship" of its primeval vegetation. Then the author of this history cites an early reference to the erection of a mill and the dismay with which it was greeted by "a cultured woman from Virginia who visited the Falls in July, 1827." In this "we ... see pictured the beginning of the decadence of the Niagara scenery...."

There is no reference to the central role the Porter family played in this decadence, centered around their ownership of the lands at the river margin, their promotion of the construction of the hydraulic canal, their pulp mill, and so forth, nor of their successful and unsuccessful attempts at developing Goat Island, and certainly not their present role as officers of the power company against which the Commission, before they were appointed to it, fought for years. When the author of this history described the commercialization of the river margin prior to 1885, he referred to a "skill bordering on military genius" where "private enterprises seized and barricaded every point of vantage commanding a visual range of the Falls...." It is difficult to envision the hodge-podge of commercial confusion of the time acting with "military genius," certainly not the Prospect Park Company with its velocipede rink, art gallery, electric fountain and other amusements, or Thomas Tugby and his Bazaar at the entrance to the Goat Island bridge. There was military genius at work, but not that of the little mill companies whose operations were condemned between 1883 and 1885.

According to the author of the commemoration, in 1869 the "movement for reclamation" of Niagara from private commercial interests, "three years before the first Federal Reservation of natural beauty was established", that is, presumably at Yellowstone (1872), "Mr. Frederick S. Church, the artist, Mr. Frederick Law Olmsted, the landscape architect, the Hon. William Dorsheimer, of Buffalo, Mr. H. H. Richardson, the architect, and others discussed measures of rescue and restoration, and during the next ten years, by means of an agitation which was never intermitted, a large body of public sentiment was created in favor of the State Reservation."

Yosemite was established as the first public land set aside to preserve its landscape by a state and federal action in 1864, as discussed above, five years before the 1869 meeting just mentioned.

The picture is given here that these men were part of a continual, persistent and coherent movement to preserve the cataracts. The agitation actually appeared to be rather casual and intermittent. More will be said about this later when details of the organized movement to create the Niagara Reservation, starting with Dufferin's conversation with Robinson in 1878 and leading to Robinson's message to the Legislature in 1879, will be examined.

The author of the history in the nineteenth annual report wrote that "the leaders of the movement," but with no names mentioned, "communicated with the Earl of Dufferin, then Governor-General of Canada" regarding the idea of creating an International Park. It seems that this communication occurred sometime in 1878, prior to Dufferin's meeting with the Governor of New York State. We are, by the continuity of ideas, to assume that these "leaders" were Church, Olmsted, Dorsheimer or Richardson. The date is given, September 26, 1878, when Dufferin introduced the idea of the International Park to the public at a meeting of the Ontario Society of Artists, in Toronto, and a bit of this speech is included in this annual report. A few weeks before this, he is said to have spoken with Governor Robinson, whom he "had the good fortune to meet," and he made his suggestion to him, an "idea which has been long present to my mind" (Dufferin, in 19 Ann Rep Comm, 1902).

Dufferin's idea was that the International Park be unadorned by the "penny arts of the landscape gardener, but carefully preserved in the picturesque condition in which it was originally laid out by the hand of nature" (Dufferin, in 19 Ann Rep Com, 1902). This sentiment appears to match that of the sentiments displayed by Olmsted and others in the 1880 Gardner report on the preservation of the scenery at Niagara Falls, the fruit of the interchange between Dufferin and Robinson, and may in itself demonstrate his actual persuasive conversations with Mr. Church.

Yet there was another party interested in the "picturesque condition" of the natural landscape to be preserved at Niagara.

Two distinct appeals were made to the Commissioners of Appraisement authorized to hear appeals from the land owners whose properties were condemned to make room for the Reservation. One appeal is somewhat bizarre, for without attribution as to the owner of the sentiment, the argument was presented to the appraisers of an appeal, the judgment against which was as follows:

"Appeal has been made to sentimental considerations. Counsel [for the appeal] has been eloquent with references to the beauties of Niagara Falls. They have spoken in fitting terms to the far-sounding roar of the cataract, the magnificence of the clouds of spray rising from its base, the rushing of the rapids, the splendor of the rainbows, and the general grandeur of the natural scenery visible from the islands and from Prospect Park. The tumultuous sound of the waters has been well likened to the "voice of God," and it was asserted that there was a certain dignity in the ownership of this property, amounting to almost a patent of nobility; and for this it was claimed that due compensation should be made."

One, or perhaps even several, of the owners wanted the State to financially compensate him, or her, for the great loss of prestige, a prestige attributable to a European nobleman by right of title, such as a baronial title, derived to the owner from such ownership. One of the owners, that is, regardless of the sentiments of the appraisers, considered himself a nobleman by reason of his or her real estate.

The Commissioners of appeal denied this request, for "the supposed dignity of ownership we can certainly see no propriety in a pecuniary award."

The following compensations were published in the Niagara Falls Gazette of July 11, 1935 (Scott & Scott, 1983), giving the names and concerns of some of the individuals who could have been involved in this appeal:

$60,200 for Tugby's Bazaar to: Thomas Tugby
$81,600 to R. F. Hill
$110,600 for Cataract House Ballroom to: Whitney, Jerauld and Company
$156,666 for Mills on Bath Island to: Niagara Falls Paper Company
$325,000 for Prospect Park to: Hans Neilson as President of Prospect Park Company
$525,000 for Goat Island, owned by:
Sarah G. Porter
Jane A. Porter
Frederica P. Burrall
Josephine M. Porter
George M. Porter
Peter A. Porter

One interest among the property owners had a profound personal investment in the primitive condition of the landscape at Niagara such that a high price might be placed on this loss of dignity. The oldest land-owners, indeed, the first landowners on the river margins at the cataracts at Niagara, after the Indian title to the land had been extinguished, were General Peter B. and Augustus A. Porter, who together owned title to Goat Island and to whose family this land was an heirloom. This was especially true as it was kept in the original condition under which it was purchased both from the State of New York, and from the Seneca Indi-ans. It was also land figuring in the treaty maps with Great Britain, a ne-gotiated settlement of which General Peter B. Porter—Peter A. Porter, Jr.'s grandfather—was an American negotiator. Indeed, as has been men-tioned before, the Porters maintained the primitive character of Goat Is-land throughout their ownership, some seventy-five years. Indeed, the Marquis of Lafayette visited the Island in 1825, the year the Porter broth-ers had issued their prospectus to develop the island. The Marquis de-scribed the land as "an aerial garden sustained by clouds and surrounded by thunder" and was disappointed that "its distance from France would not permit him to purchase it as it would make a delightful residence" (Lafayette, in Porter, 1900).

To return to the thread of the 1903 history of the movement to preserve the Niagara landscape, the third party, after Lord Dufferin and

Olmsted and James Gardner, with a strong interest in preserving the natural scenery, and restoring it, were the former owners. After losing their land at the cataracts, the Porter family went on to develop the Niagara Falls Power Company on land they owned just upriver of the Reservation. This company and others like it were opposed by Green and the Commissioners constituting the board before the members of it, excepting Green and George Raines, were replaced in 1898. The exception of Green himself, a staunch and irreconcilable advocate of the landscape objectives of the Commission declared by the laws creating the Commission, was notable. With the inclusion of the Porters and their interests on the Commission, two objectives could be pursued, protection of the primitive landscape, and pursuing the economic investments of the family and allied interests. The Porter owners had lost Goat Island in 1883, but by 1900, they were again largely in charge of its administration.

Whether Green knew or not the relationship between his colleagues on the Commission and their role in the commercial interests threatening to overwhelm the integrity of the cataracts the Commission was appointed to protect, will be discussed separately.

The author of the 19th annual report of the Commissioners continued his history of the movement to establish the Reservation. Governor Robinson went directly to the New York State Legislature, in 1879, within a relatively short time after meeting with Lord Dufferin, and requested a study to investigate the matter at Niagara. He had it passed through the Legislature with alacrity. Grover Cleveland would become Governor of the State in November of 1882, so the passage of the 1879 bill was independent of his influence, that is, not associated with a reform platform.

Under the historic section "obstacles overcome" in establishing the Reservation, the first obstacle was overcoming the lack of precedent for voting public money for the purchase of land by eminent domain in order to preserve its beauty. The second obstacle is curious and there are two elements to it.

The *first* is the intent to create a "park" defined as one "laid out with neatly trimmed lawns, formal pathways, geometrical flowerbeds, composition statuary and cast-iron benches," a concept much embraced by the Canadian Commission, in spite of Lord Dufferin's interests in preserving and restoring the ecosystem in the area in the vicinity of the cataracts. It is curious that part of the concept of "artificial embellishments of decorative landscape gardening" is the notion of "laying out" the area of the Reservation with "conventional designs of paved roadways." One wonders how the roadway conventions differed from those in place and intended to be laid in the Reservation.

The *second* is more interesting for it involves early deliberations on the Reservation's boundaries.

"People ... imagined that it was proposed to take a vast tract on each side of the Niagara river, extending from far above the Falls to below the Whirlpool ...." One may well imagine the alarm in the minds of developers of land along the upper river at this prospect. Condemning land just north of the cataracts would eliminate the Schoellkopf Milling District at great cost to the State, so that was understandably, perhaps, abandoned. But there was no reason to believe that the taking lands along the upper river would be beyond the grasp or the best interests of the State. Indeed, this land was cheap, or so the Niagara Falls Power Company found it when they bought up the shoreline for their plants, railroad rights of way, workmen's settlement and industrial dependents. In 1887 and in the years previous, land above the falls was relatively cheap. When it came out that a group was negotiating the capitalization of an extensive power development corporation along the upper river in 1887, a cable to that effect was apprehended by a telegraph operator who, with an associate, "had ridden the surrounding territory and secured options upon a great deal of land, so that when the quartet of promoters attempted to secure land for the company and themselves, 'the cat was out of the bag' and the prices of all lands anywhere in the vicinity of Niagara Falls had advanced greatly, never to come down to the former value .... As a result, the tunnel company and its discreet promoters paid comparatively high prices for lands purchased thereafter" (Adams, 1927).

Extensive property just upriver from the Reservation in 1853 was owned by the trustees of Samuel DeVeaux's estate, managing Lot Numbers 46, 48, 49, 50 and 52 of the Mile Reserve immediately upriver from Lot Number 44, the 100 acres of the Stedman Farm sold to Peter B. Porter in or prior to 1805, a parcel out of which was purchased by Mr. DeVeaux (Loker, 1963).

By 1903, "the landed estate [of the companies associated with the Niagara Falls Power Company] is located mainly within the municipal limits of the present [1927] City of Niagara Falls. About one-third is located on the river front, with projected wharfs and dock facilities, of nearly 2 miles along the navigable channel" (Adams, 1927).

Another objection to the course of establishing the Reservation, conversely, was that the proposed limits were too modest. For the author of this essay in 1902, "Public opinion has grown more liberal on the subject during the past twenty years." The author resorted to an anachronism by stating that the Commissioners of 1902 (the "we" of "we are convinced,") were of the opinion that the boundaries were appropriate in 1883, without stating what their opinion was in 1902, and whether it had become liberalized along with public opinion.

The author of this history then proceeds to detail the accomplishments of prior Commissions. Prominent place is given to the offer by Albert H. Porter to donate valuable land adjacent to the Reservation in 1887 (4 Ann Rep Comm, 1888) for a "Museum of Natural Science and History of Niagara" and the Legislature's failure to take him up on it. The present Commission congratulated themselves that since their appointment "the most extensive and substantial improvements have been made within the last five years, during which period the present Commissioners have consummated the well-considered plans of their predecessors and have effected many more improvements in harmony with the original conception of the Reservation."

Since the appointment of this particular Commission, the link with the parkmen of New York City was broken. The link with Olmsted and Vaux's colleagues in landscape architecture, Downing Vaux and associates, was passed over for consultation with the State Architect and State Engineers. Landscape consultation with Central Park and New York City parkmen such as Parsons, Wirth and W. S. Edgerton was terminated, it seems, as well. These Commissioners appear to have decided that the "well-considered plans of their predecessors" were "consummated" and yet, the reason that Parsons had been elected Landscape Architect to the Commission in the year immediately before the appointment of the new Commission in 1898, where his name appears upon the masthead of Commissioners and Commission officers (14 Ann Rep Comm, 1898, for 1897), was that the work of restoration was only commencing.

Note also that the shoreline on the plan of Olmsted and Vaux had been redesigned by Welch, filling in the canals cut into the shore, but curiously leaving many of the extensions into the river, such as wing-dams and races, such as the race or excavation "creating" so-called Willow Island, intact. He had even created a "Port Day Park" adjacent to the inlet canal of Schoellkopf's hydraulic canal which formed the east boundary of the Reservation.

Before 1898, the Commission had been consulting "old pictures, the recollections of old residents, and careful observations of the natural environment" in their development of a restoration protocol. "Unsightly embankments were demolished and the mutilated shore restored to its normal condition" except that Mr. Welch had been using fill from power company and hydraulic canal excavations to bury the old shoreline. It is noteworthy that the concern of the old Commissions was only to grade the earth "where imperatively necessary," in keeping with protection of the old grade, or grades, that is the topography in existence at the Reservation's creation. This policy is to be compared to that of the first decade of the twentieth century where extensive and intensive grading was employed adjacent to all structures, including roads, and throughout all of Prospect Park, largely destroying the land's original character and vegetation.

"Denuded places were planted with trees, care being taken to use only indigenous varieties." This practice, in the coming decade, was also to change.

In 1900 the wooden stairs down to Luna Island, which had preserved much of the vegetated character of this botanically striking area, was removed and its natural character stripped during the construction of an imposing and graded stone stairs. The same fate would befall the bluff overlooking the Terrapin Rocks, where the vegetated elevation would be completely denuded, its old wooden stairs removed and a graded slope installed in its place. Both unnatural constructions were described by the Commissioners of 1902 as "more natural."

The Luna Island stairs "was the beginning of the class of better and permanent structures which the Reservation sadly needed." Thus do the Commissioners declare their break with the previous twenty years administration which had built the stone bridges and two structures: the building at the base of the inclined railway in Prospect Park and the visitor's shelter at the entrance to Goat Island. The Commission of 1898 takes the credit, in this history, for Downing Vaux's bridge to the First Sister and the alteration of Prospect Point to restore its natural appearance, both of which were the outcome of the previous Commission's design and study. The "transformation" of an area in Prospect Park called "Hennepin's View" provided a convenient area for the deposition of fill, "earth and rock," creating a hill affording an improved view from that point, without its previous "wooden platform and stairs" by which visitors had previously improved their vantage. It had been stated a previous annual report that Downing Vaux had made detailed plans for this area (16 Ann Rep Comm, 1900: "$147.50, Dec. 1899, to Vaux & Emory, Hennepin's view and buildings").

This destruction of the original aspect of Hennepin's View will be typical of the obliteration of the original topography and ecosystem intended for preservation in the terms of this and future Commissions. Structures which had been built in natural areas, such as the bluffs over-looking Luna Island and Terrapin Rocks, areas seeming to demand elevated structures for views, under previous administrations had tolerated wooden structures built into the landscape. Structures may have been conspicuous, but their image provided the contrast necessary to impose on the minds of the visitor the clear separation of the manmade and the natural. Subsequent administrations, after the power company interests took over the Commission, would alter the landscape they were meant to preserve, by the accommodation of conduits for sewage, telephones, water, electricity, roads, parking areas, etc., with earthwork modifications giving a result that was far from natural, but gave the appearance of something natural. A dirt slope may not appear as artificial as a wooden stair, but the wooden stair preserved the natural setting with its vegetation, while the dirt slope, or dirt slope in which stone stairs were embedded eradicated it.

Faking nature — The notion of faking nature while destroying it begins with the Commission of 1898. The area of Hennepin's View had been buried beneath a pile of "earth and rock," but it was made to resemble "as closely as possible a natural formation."

The phrase "new and desirable views" began to take on a new significance. In imitation of the language of the landscape architects Olmsted and Vaux, the Commission began to engineer views on their own. In addition to the fill used to make a mound out of Hennepin's View, at a point a little north of it, at the north boundary of the Reservation, "the point was raised; the high bank rip-rapped with large rustic stones; and a new and desirable view opened to visitors." One wonders whether a bit more material excavated in power plant activities was gotten rid of in this section of the Reservation as was being done all along the mainland shoreline of the Reservation.

The 1902 Commissioners boast of their control of "money-making enterprises" which were being constantly proposed for inclusion on Reservation grounds, but which the Commissioners "resolutely resisted" from the beginning. However, this resistance was continual until 1898 and the capitulation to the railroad company which wanted to operate in the Reservation. The company had wanted to set up facilities in the Reservation for a long time, but the Commission had hitherto succeeded in blocking them. Even though still President of the Commission in 1898,

A. H. Green had opposed this development, and had refused to sign his name to the contract later drawn up between the Commission and Caryl Ely, President of the Niagara Falls and Suspension Bridge Railway Company.

Another instance of the new Commission acting independently of Green, was their voting in November, 1898, to rename Bath Island, Green Island after him. That such a decision may have jeopardized Green's credibility as a disinterested President of the Commission, may have been deliberate, and in any case it appears inappropriate while the President was still incumbent. This little gift of the junior Commissioners to their chief was intended "to pay a tribute of public respect to their distinguished colleague." "For twenty years as a Commissioner and for the past fifteen years as President, he has been a most zealous and efficient conservator of the public interests centered in the State Reservation at Niagara; and the application of his name to the island was not only most appropriate to the object, but also a merited tribute to a fearless and faithful public servant."

Green was indeed fearless and faithful. His withdrawal from the Commission in the ways he had previously dominated it attests to the faith he had in the Reservation as envisioned by its founders. That he was fearless is perhaps indicated by the powerful interests against which he was fighting. Green may have been away in New York City, but he was not inactive. He was, in fact, appealing to the federal government for control of diversion rights. The other Commissioners had re-elected Green President when the new Commission formed in 1898, as discussed above, even though a local newspaper expected him to become "former President." The other Commissioners, to maintain their own credibility as Commissioners, could not give Green any idea of their personal interests in issues at variance with Green's.

That the new Commission was effective in appropriating money from the New York State Legislature is catalogued in the 19th Annual Report as follows:

Pre-1898 appropriations:

By chapter 302, Laws of 1891 $15,000
By chapter 356, Laws of 1892 $15,000
By chapter 726, Laws of 1893 $25,000
By chapter 358, Laws of 1894 $20,000
By chapter 932, Laws of 1895 $20,000
By chapter 950, Laws of 1896 $10,000
By chapter 790, Laws of 1897 $15,000

Post-1898 appropriations:
By chapter 606, Laws of 1898 $15,000
By chapter 569, Laws of 1899: $30,000
By chapter 570, Laws of 1899: $15,000
1899 total $45,000
By chapter 419, Laws of 1900: $51,000
By chapter 420, Laws of 1900: $77,000
1900 total $128,000
By chapter 645, Laws of 1901 $44,665.

The New York State Legislature was obviously very helpful in assisting in the Reservation's preparation for the 1901 Pan-American Exposition.

In a convincing discussion of how, by saving the public the fees they would have normally incurred while visiting the Reservation before that land became State property, the Reservation had paid for itself four times over, the author of the Reservation's history displayed a fine understanding of how the Reservation's twenty-year history "has afforded a practical demonstration in this country ... that the preservation of the beautiful, the picturesque and the historic 'pays.'" If the Commissioners themselves were local businessmen with large financial commitments to the kinds of economic interests dependent on the tourist trade, they would be in a special position to appreciate the Reservation's usefulness, in fact, they may have seen this before the Reservation had yet been conceived in the political arena.

Again, a "scenic driveway" is proposed, to run from the cataracts to Fort Niagara on Lake Ontario, some 16 miles, and the State is asked to consider installing this route. The Commission for decades had been considering this idea, even if only as far north of the Reservation as the Whirlpool. The scenic route discussed in this report is of historic and scenic value, commemorating the various war-associated events along the lower Niagara River.

A change in attitude — The intense dislike and distrust of the aggrandizing power companies that had characterized the Commission reports of the years previous to 1989 had disappeared abruptly from the annual reports to the Legislature. Indeed, there appears to have been a kind of reconciliation with the power development interests in small concessions that the Commission, before 1898, had looked upon with contempt: the issue of using stony material from corporate excavations adjacent to the Reservation as fill, and the law that entitled the Reservation free water and electricity from the Niagara Falls Power Company in return for diverting water from the River.

Now the history of the Reservation in the 19th report turns to the issue of the "endangerment" of the Falls due to unregulated water diversions. Note that at least two trustees of the Niagara Falls Power Company are, at this time, officials of the Commission, one of which, Peter A. Porter, Jr., is a historian and an author.

The text of the history of the Commission attitude toward power development strove to make no distinction between the present Commission, and the attitudes of past Commissions. The author of this history recognized the past Commission's uncompromising rejection of development schemes.

Reference is made to the 1889 bill to authorize the Niagara Hydraulic Electric Company "to erect machinery under Niagara Falls for the purpose of utilizing the power of said Falls for manufacturing electricity." This was the company incorporated in 1887 in Virginia which, even after having been "authorized capital stock of $20,000,000," (Adams, 1927) did not proceed into development.

That Mr. Peter A. Porter, Jr. had the interests of the Niagara Falls Power Company at heart is indicated by his visit to Buffalo where he convinced promoters of the Virginia company in Buffalo to drop their support because the project he and his associates envisioned "was far more promising of practical success" (Adams, 1927). Adams made no further reference to this company, yet their promoters persisted. In May 30, 1889, the Virginia company approached the Commissioners of the Queen Victoria Park, Ontario, and were granted by that Commission, with the approval of the government, a franchise to develop power in the park. "Owing largely to disputes between the promoters and the financial interests the Niagara Hydraulic Electrical Company failed to obtain the $50,000.000 deposit required to pay the first two years' rental and as a result the negotiations were terminated in December, 1889 (Way, 1946). When the Canadian Commission granted exclusive rights to develop power in the park to interests associated with the Niagara Falls Power Company in 1892, the Virginia company disappeared from local history, at least under the name used by historians.

At any rate, the Commissioners of the Niagara Reservation, until 1898, vigorously opposed diversions in the upper river because they would not "countenance any scheme the success of which would be likely to result in the defacement of the landscape or in any way interfere with the performance of the duty entrusted to them, namely, that of restoring the scenery to its natural conditions" (quote used in 19 Ann Rep Comm, 1903).

The author of the nineteenth annual report then proceeded to write that "defeated in the attempt to erect power works immediately under the Falls, enterprising promoters soon devised a plan of evading the uncompromising position of the Reservation Commissioners, by taking water from the Niagara river above the Reservation and conducting it by subterranean tunnel to the river below the Falls, utilizing its power in transit."

These promoters were, among others, Peter A. Porter, Jr. himself, and Alexander Porter of the Niagara Commission. By Mr. Porter's own history published in the 16th Annual Report, the plan had been to build a tunnel under Goat Island in 1877. It was the legislation of 1879 in New York that precluded the ability to use Goat Island for generating power, and not the Commission of the Niagara Reservation, which did not exist at that time (Gardner, 1979).

The essay acknowledged Green's personal role in approaching the federal government for regulation of diversion rates at Niagara Falls. Note that Green was doing this solely to preserve the natural environment, or the spectacle and experience of it, in the vicinity of the cataracts. Note that the President of the 1903 Commission of the Niagara Reservation was pursuing this—no mention is made that the rest of the Commissioners were backing Green's initiative, nor that these same men were continuing the anti-diversion initiatives of previous Commissions.

Diversion — Events in the Great Lakes "led the President of this Commission to urge upon Congress the creation of a United States Commission ...." The other Commissioners did agree, perhaps, according to the author of the historic essay, that "great good cannot fail to come from an intelligent and thorough investigation of this important and far-reaching question." However, in this report, the other Commission members remained otherwise mute on the subject of diversion.

The treatment of the power issues concerning the Reservations on either side of the river by the Commission of the Niagara Reservation in this 19th annual report to the New York State Legislature must be seen as an attempt to declare the disinterest of the officers of the American Commission in corporate policies of the power companies. Knowing, however, that two of the officers of the American Commission were directors of the Niagara Falls Power Company itself, lends a degree of incredibility to such an effort, and one is rather forced to conclude that these officers had managed to keep their business ties secret from the public. Additional evidence of the secrecy of the commercial interests of these officers was the fact that there were not only two directors of the company on the Commission, but, in fact, there were three. This third power company director had been an officer of the Commission since May 22, 1885 (2 Ann Rep Comm, 1886). This third director will be discussed below.

Examples in the nineteenth annual report where the American Commission includes the interests of the business interests of certain of its members in their criticisms of cataract and park deterioration must be seen as a demonstration of the conflict of interests involved. Criticism of their own commercial interests while apparently pursuing the interests of the public must then be an attempt to give evidence for their impartiality.

Nowhere in the report is there an attempt made to have the executives of the power companies share responsibility for the deterioration of the cataracts and adjacent public lands by their plans to industrialize areas set aside, or reserved, for public use.

Ontario — The author of the historic essay now continues the story of the Queen Victoria Niagara Falls Park and continues its history, emphasizing its mandate to create such a park, called, in the essay, the "Niagara Reservation on the Canadian side of the river," recalling the "zealous advocacy, by the Earl of Dufferin, of the proposal to rescue the Falls of Niagara from vandalism." A quote is made from a Canadian document written in language emphasizing the natural beauties of the land to be included in the Canadian park. The Canadian Commissioners, not the original ones in all cases, at the time of the essay were identified as the Hon. J. W. Langmuir, Chairman; the Hon. George H. Wilkes, the Hon. James Bampfield, and the Hon. A. W. Campbell. No indiscretion was employed in naming the Canadian Commissioners, while the executives of the power companies remain anonymous.

In contrast with the avowed aims of the Canadian Commission, the next section of this essay is entitled "Disfigurement of the Canadian Park" and a photograph of the destruction at the flank of the Horseshoe Falls is shown, November, 1902, by the Canadian Niagara Power Company, actually a subsidiary of the Niagara Falls Power Company.

"This" American Commission regrets the decision of past New York Legislatures "in permitting the diversion of water from the Falls" by American capitalists, but an even greater regret is felt toward the Ontario Legislature, which has "not only granted the right to subtract a large volume of water from the Falls, but it has also permitted the power companies to invade the Victoria Park and erect their unsightly structures in full view of both the American and Canadian Reservations." Just as in the case of awards by the New York State Legislature, the amount of water granted for diversion by the Ontario government was unlimited.

The Canadian Niagara Falls Power Company would continue to disfigure the Canadian park, with the Canadian Park Commission's consent, destroying Cedar Island, a large, wooded island at the head of the Horseshoe Falls, for the sake of the company's diversion capability (Tiplin, 1988) and destroying part of the botanically important wooded ravine (Bowman's Ravine) at the Whirlpool by using it as a dump for excavations for the plant's wheelpit (Seibel, 1985).

But perhaps the most serious event of the year (1902) was the death of two Niagara Reservation, New York, workmen on October 28, 1902, from debris discharged across the river from dynamite operations during the course of power-house construction by the Ontario Power Company in the Canadian Reservation, constructions and their manipulation of the "contour of the shore" which had "been changed noticeably and given an artificial appearance. At the Horseshoe Fall, the shore line has been shortened about 200 feet. The diversion of water by wing-dams has made dry a large area of the river bed in the neighborhood of the Dufferin Islands. And the power houses will present incongruous objects in a landscape which was taken to be preserved as nearly as possible in its natural condition."

In July, 1902, the "Commissioners of the State Reservation" "requested the Canadian Commission to take no final action" to allow the development of both power companies in their park, although, if a request was made in writing, it was not included in the 19th annual report for that year, and consequently, the signatures indicating unanimity among the members of the American Commission cannot be confirmed.

"On August 2d, Commissioner [Alexander] Porter and Superintendent Welch met with the Canadian Commissioners at their office in the Queen Victoria Niagara Falls Park." There, they protested to the Canadian Commissioners that "some power company or user, was to be granted or had been granted the right to construct penstocks over the high bank at a point about where the old "Table Rock House" was located, also for the erection of a power house on the lower slope, under the bank, at the same point."

This was essentially a complaint against the Ontario Power Company, a rival American-capitalized power plant to the Canadian subsidiary of the Niagara Falls Power Company, the Canadian Niagara Power Company. The Ontario Power Company was building its power plant at the base of the Horseshoe Falls in the direct view from the prospect points at the western boundary of the Niagara Reservation in New York. The transformer house of the rival plant was located on the high bluff above the Queen Victoria Niagara Falls Park directly west of the power-house at the base of the gorge (see historic map, 1907 opposite p. 36, Seibel, 1985). It does not appear to have been a complaint against the Canadian Niagara Power Company, whose interests were represented by Alexander and Peter A. Porter, Jr., and one other officer of the American Commission, whether the Canadian Commissioners knew it or not. We are also expected to believe that the Porters were unaware of the Ontario Power Company's local backers: "some power company or user, was to be granted or had been granted the right ...."

It should be born in mind that in 1899, a year after the appointments to the Niagara Reservation in New York, the Ontario Legislature (62 Vict. 2 cap. III, par. 35) voted to deny the Canadian subsidiary of the Niagara Falls Power Company the monopoly which had been included in the original franchise granted the American company in 1892 (Way, 1946, p. 65). "By this time [1899] requests for franchises to generate power had been received from other companies and it was generally felt that the monopoly clause should be canceled" (Way, 1946, p. 66).

As early as 1887, an entity called the Ontario Power Company had received a charter from the Federal Government of Canada, but had been blocked by the monopoly clause granted the Niagara Falls Power Company interests in 1892. "As soon, however, as the exclusive rights were annulled, the Ontario Power Company again applied for a franchise and in April 1900 an agreement was obtained" (Way, 1946, p. 67). This company had originally intended to take its diversion from the Welland Canal, but a second agreement of June 1902 "in consideration of an increased rental [to the Canadian Commission] was permitted to tap the Niagara River at the Dufferin Islands and by means of underground constructions develop electricity in a power house situated in the gorge immediately below the Falls" (Way, 1946, p. 68).

The way the Niagara Falls Power Company had interacted with the appearance cataract environment so far, on both sides of the river, was to build its plants according to the design of one of the finest architectural firms in the United States (McKim, Mead & White), and to establish its plants in places along the Niagara River that would not compete with the view. It was in their best interests, therefore, to protest the construction of rival plants that were an aesthetic detriment to the Niagara environment, plants and operations that were "a blemish and defacement."

The American delegation to the Canadian Commission emphasized their twenty-year campaign of removal of all unsightly commercial structures within the boundaries of the American Reservation. They excused the, by that time, grotesque proliferation of commercial structures of the Schoellkopf Milling District and power generating plants just north of the Reservation on the Gorge Wall in New York because the cost of including the district in 1879 would have been prohibitive. Perhaps the same argument could have been made by the Canadian Commission in terminating the north boundary of the Queen Victoria Park to leave private the grand Clifton Hotel and the entrance to the carriage bridge in view of the falls.

Thus, if the monopoly of the Niagara Falls Power Company had not been broken, the visual integrity of the Canadian park may, however, have persisted.

The American Commission had even, in 1902, removed their office building from Green Island "with the understanding that it was the common policy of the Commissioners of the Queen Victoria Niagara Falls Park, and of the Commissioners of the State Reservation at Niagara, to remove all unnecessary and incongruous structures from the immediate vicinity of the Falls, in order to restore the Falls and their surroundings, as nearly as possible, to a state of nature." That year, the Superintendent also removed the cottage dwelling house at the entrance to the bridge to Bath Island, on Goat Island and a gate house "adjacent to the entrance to the steel arch bridge" (Welch, in 19 Ann Rep Comm, 1903).

The reply by the Canadian Commission was to refer to the economic strategy established for the operation of the Canadian Reservation, a strategy that may have been calculated from the beginning to require the Commission to introduce private utility plants (water, electric, transportation) into their park to generate revenue in order for the Commission to function. The Americans could always rely on tax revenue from the State, but in Ontario, legislation had made it nearly impossible to use public revenue.

"With this condition" replied the Canadian Commission, "it made it necessary to deal with the several electric railways and power companies in a different manner from that which would be possible were the Commission in an absolutely independent state of financial resource" as apparently the American Commission was seen to be.

To counter the "defacement" accusation, the Canadian Commission assured Porter and Welch that the power plants would "be in form as to style of architecture, so as to be in the least degree objectionable."

On September 17, 1902, at a meeting of the American Commission on September 18, 1902, Green was asked to write a letter to the Canadian Commission, which he apparently did, the text of which was published with the 19th annual report, essentially reiterating the objections made by the American delegation of A. Porter and T. Welch.

A photograph of the American channel bereft of water due to ice jams in the upper river (March 22, 1903), showing, perhaps the potential future appearance of the American channel should diversion continue unchecked, was included with the publication of the letter.

The Commissioners end their report for 1902 with a statement of their policy, tucked away after a list of the publications which had been included in previous annual reports. That policy "has been to restore and protect nature; to surround the cataract with conditions that would conduce to the freest and most unqualified enjoyment of the great spectacle and to stand like an immovable bulwark between the highest interests of the State and any attempts to divert the Reservation from the purposes for which it was created. How far the Commission has attained the mark of its ambition, it leaves the public to judge." This brief, two paragraph reiteration of the Commissioner's policy was signed by all the Commissioners, including Green.

The Superintendent's Report — The platform "of the lower terminal station of the Inclined Railway" designed by Downing Vaux, had been destroyed by falling ice during the winter of 1902-1903 and was replaced (Welch, in 19 Ann Rep Comm, 1903). The State Architect was consulted for modifications in the upper terminus of the same railway.

The Commission continued to proceed to avail itself of "free" electricity from the Niagara Falls Power Company according to the law granting such a boon to the Reservation. The power had previously been purchased from the Schoellkopf electric plant, according to disbursements by the Commission treasurers. Although the power was free, the electrical infrastructure on the Reservation was not, and a requisition from the Legislature had to be made.

Two electrical engineers were named by Welch, Paul M. Lincoln and Harold W. Buck, "concerning the most desirable system of electrical installation for lighting the Reservation ground and buildings." Buck had apparently submitted a plan to the Board on a previous year and appropriations made. The Legislature had already appropriated $7,000 for the installation (chapter 645 of the Laws of 1901), but an additional study concluded that an additional appropriation be made.

What Welch did not tell the Legislature was that both men were not disinterested engineers, but were employees of the Niagara Falls Power Company. Buck was employed between 1900 and 1907, the engineer in charge of Power-house Number Two and "of the initial half of the Canadian Niagara Power Company development as well as all other portions of the electrical system" (Adams, 1927, Vol. 1, pp. 285–286). He was also administrator of the "electrical engineering department" (Vol. 11, p. 353). Mr. Lincoln would be in charge of constructing the electric transmission lines to Buffalo from the Niagara Falls plant and he worked on the electrical construction staff between 1895 and 1902.

The Pan-American Exposition was over and "four temporary carriage stands have been established on the Riverway, between Niagara street and Bridge street" and more normal competitive relationships were resumed.

The Conclusion of the 19th Annual Report — At the very end of the 19th annual report, after two essays, one historical, one geological, are appended "Documents relating to the International Commission to investigate and report upon the conditions and uses of the waters adjacent to the boundary lines between the United States and Canada" and the florid address of James C. Carter, Esq. "at the Dedication of the State Reservation at Niagara, July 15, 1885."

Andrew H. Green had prepared a document for Hon. Thomas C. Platt, Senator of New York, to introduce in the United States Senate as a joint resolution. Platt did so on February 6, 1902 (Senate Resolution 52). The resolution authorized the President of the United States "to invite the government of Canada to join in the formation of an *international commission to examine and report upon the diversion of the waters that are the boundaries of the two countries.*"

The commission, as adopted under the River and Harbor Appropriation bill, section four, was to be appointed by the President of the United States and his counterpart in Canada. There were to be three members each for a panel of six. The American members would be "one officer of the Corps of Engineers of the United States Army, one civil engineer well versed in the hydraulics of the Great Lakes, and one lawyer of experience in questions of international and riparian law" in addition to such expert staff as these individuals deemed necessary to resolve specific questions raised between the two countries. The Commission would form in response to a question to which they would apply themselves, and disband when a formal resolution of the question was reached.

The Great Lakes region is the area to be dealt with by the International Commission, and the "conditions and uses of the waters" in the Lakes, the "maintenance and regulation of suitable levels; and also upon the effect upon the shores of these waters and the structures thereon, and upon the interests of navigation..."

This very Commission would also explore an issue raised under the federal administration of President McKinley, who was assassinated in Buffalo at the Pan-American Exposition after visiting the Niagara Reservation:

"The said commissioners shall report upon the advisability of locating a dam at the outlet of Lake Erie, with a view to determining whether such dam will benefit navigation ...."

There is no specific reference to the diversion of Great Lakes water to serve or adversely affect the interests of the utilities, especially those situated at Niagara Falls, and their future interests in the subsidiaries proliferating in the Canadian park.

The River and Harbor Bill became law "by the approval of the Executive June 13, 1902," that is, of the new office of Vice-President Theodore Roosevelt, inaugurated into the office of the Presidency at the death of President McKinley on September 14, 1901. Roosevelt, a reform Republican, would become known in history for the breaking of the great trusts, the monopolies of banking and other financial institutions—especially the monolithic financial empire of J. Piermont Morgan, one of the investors in the Niagara Falls Power Company. With Roosevelt would come federal legislation acknowledging the limits of, and legislating the conservation of, the natural forest and mineral wealth of the United States.

20th Annual Report of the Commissioners of the State Reservation at Niagara for the fiscal year from October 1, 1902, to September 30, 1903. Transmitted to the New York State Legislature February 4, **1904**.

The Deaths of Andrew Haswell Green and Thomas Vincent Welch — On February 13, 1903, the 19th annual report was transmitted to the New York State Legislature. On March 30, 1903, the incumbent Commissioners were re-appointed by Governor Odell. George Raines, Thomas P. Kingsford, Charles Dow and Alexander J. Porter in addition to Mr. Green were to serve the next five-year term. Later in the year, Alvah K.

Potter of Lockport was appointed by the Governor to replace Green, and Charles M. Dow of Jamestown was elected President.

The Commission met in Niagara Falls, on December 22, 1903. The previous July, Peter A. Porter, Jr. resigned his position as Secretary-Treasurer "to take effect October 1 following." Thomas V. Welch, who had served as Superintendent since the founding of the Reservation, was elected Secretary, and Edward H. Perry, who apparently had been serving as an Assistant Superintendent, was elected Treasurer.

Mr. Welch had died of typhoid fever on October 20 of that year, and Perry was appointed Superintendent and Secretary in his place. Twenty-four days later, on November 13, 1903, Andrew Haswell Green was dead, killed by an assassin (Rubbinaccio, 2013) on his doorstep outside his home on Park Avenue in New York City.

Two memorials were prepared by the Commissioners for each of these men, and some account is made of the accomplishments of each. A memorial was addressed by the Commission to the family of Mr. Green, briefly eulogizing on Green's life. A similar memorial was addressed by the Commission to the family of Mr. Welch.

According to his memorial, Thomas Welch had many civic involvements in the City of Niagara Falls:

"In addition to his services to the State in the Assembly and as Superintendent of the State Reservation at Niagara, he held many positions of trust in the village and city of Niagara Falls. As village Clerk and Trustee, Chairman of the Board of Supervisors ... member of the Board of Education, Trustee of the Niagara Falls Public Library, Trustee of the Niagara University, member of the Soldiers' Monument Commission, Secretary of the Memorial Hospital Association, President of the Civic Club, President of the Niagara County Pioneers' Association, Vice-President of the Niagara Frontier Historical Society, member of the Niagara Frontier Landmarks Association, member of St. Mary's Roman Catholic Church ... organizer of the Niagara County Savings Bank, Director of the Niagara Falls Power Company, and in numerous other relations, official, business and philanthropic, he was identified with practically every public enterprise in Niagara Falls for the past 25 years."

In all of the literature reviewed in the present study, only two references have occurred to indicate Mr. Welch's relationship with the Niagara Falls Power Company: buried away in the list of civic organizations mentioned above, and in the history of that company written by Dean Adams nearly a quarter of a century later (Adams, 1927).

When Mr. Welch and Alexander J. Porter, also a director, went to the Canadian Commission the previous year to remonstrate against that board's permitting the Ontario Power Company to build in the Queen Victoria Park, they had gone representing the Commission of the State Reservation at Niagara. They were also representatives of a rival power company that had once been promised a monopoly by the Canadian Commission, but by whose actions the monopoly was broken. The Ontario Power Company was permitted instead to build a power plant. Mr. Welch had served as Superintendent of the Reservation throughout the years when the Commission protested the charters of the power companies by the New York State Legislature. He had given "18 years of faithful and intelligent devotion to the execution of the Commissioner's plans for the development of the State's famous scenic park."

In 1898, Alexander J. Porter had been appointed Commissioner of the State Reservation. A year later Peter A. Porter, Jr., a power company director, had been made Secretary-Treasurer. The Superintendent had been elected a Reservation officer as early as 1885. How long had Welch been a director of the Niagara Falls Power Company and what was his connection with the other interests of that company, particularly in transportation issues leading to the virtual monopoly of public transportation and tourist-related services on both sides of the border by the International Railway Company?

It was a tragedy that Welch had succumbed to the contaminated water supply of the City of Niagara Falls at a time when water conduits were being installed in the Reservation, and the adjacent City was actively engaged in purifying its water supply.

As to the Commission's business of that year, there was no hesitation on the part of the new Commission to require an additional appropriation of $12,000 from the Legislature above the appropriations of the previous years for equipment to supply electric light and power, at no cost, to the Reservation—plus an additional $3,500 for converting the Inclined Railway from water power to electricity. Two estimates were provided by Mr. H. W. Buck, Electrical Engineer, but still no reference is made by the Commissioners or by Mr. Buck as to Buck's relationship with the Niagara Falls Power Company, as noted above. Electricity was still being bought from the Schoellkopf plant, the Niagara Falls Hydraulic Power & Manufacturing Company, the disbursements to which were tallied in the budget of the 20th annual report.

Edward H. Perry, the new Superintendent, referred to the continued danger of dynamiting by the Ontario Power Company which had to be "cautioned" several times during the year.

Memorial Addresses: Thomas Welch — A public meeting was held in the City of Niagara Falls to honor the memory of Hon. Thomas Vincent Welch. The various addresses given provide additional information on Welch's relationships with colleagues and peers in that City as late as 1903.

Mr. Caryl Ely, President of the Niagara Falls and Suspension Bridge Railway Company as discussed in earlier annual reports, was a colleague of Welch not only in the Assembly of the New York State Legislature, but he and Welch and three other men "in February, 1886, met at the office of the superintendent of the state reservation on Bath Island, and discussed the practicability of Mr. Evershed's plan and its possibilities" (Adams, 1927). Within eight months of the Reservation's existence, the alliances that were to result in the formation and development of the Niagara Falls Power Company were created. It was Welch and Ely that took a bill "to procure a charter by special act of the legislature" drafted by Ely, to Albany, where it was "submitted to Peter A. Porter ... then member of the Assembly from the Niagara Falls District" (Adams, 1927). Careful note must be made to distinguish between Peter A. Porter, the son of Gen. Peter B. Porter, and Peter A. Porter, Jr.—which Adams failed to do. The senior Porter and son of the general had died during the Civil War. The "member of the Assembly" indicated by Adams, was the same gentleman who served as Secretary-Treasurer to the Commission of the State Reservation, as of Feb. 1, 1900 (19 Ann Rep Comm, 1903). Welch and Ely were together officers and directors of an early organization of the Niagara Falls Power Company and were to continue their association throughout Welch's term as Superintendent.

In the address given by Charles E. Cromley, Mr. Welch's relationship with the Niagara Falls Power Company is obscured. In reference to Welch, without mentioning corporate names: "the great corporation whose establishment here has caused such marvelous growth and material prosperity of our city found in our dead friend one of its first champions. With a few others he labored in and out of season, early and late, to bring this vast combination of capital to our city ...."

Mr. Ely related he first met Welch January 2, 1883 in the State Assembly. Ely is said to have resided in Buffalo at the time of the address, but when he worked with Welch to create the Niagara Falls Power Company, he had lived in Niagara Falls (Adams, 1927). Ely ended his address unambiguously, that he had recollections of Welch, that "over and above the recollection of his public services there comes to me, begotten of 20 years of intimate association in public, private and business life, the memory of the manifold virtues of the man." This is completely plausible because they had spent two decades together working toward the same commercial ends.

Green's Last Speech — Included in the twentieth annual report is a speech said to have been written by Mr. Green, but not delivered by him. There are many elements of the speech that appear to be characteristic of Green's thought, but others that do not. One may wonder if the speech was tampered with before its delivery as an extension of some of the profound conflicts of private and public interest detailed in the analysis of the first twenty annual reports above, and in examples of other events to be examined in subsequent chapters below.

The implications of such a suggestion, that the speech was altered before delivery, are grave because it might logically follow that Green might not be alive in 1904 to protest upon seeing the speech as transmitted to the New York State Legislature within the first three months of that year.

The address was to have been delivered in Niagara Falls by Mr. Green at a meeting of the American Park and Outdoor Art Association, July 7, 1903, but "At the last moment, Mr. Green was unavoidably prevented from attending the convention in person, but sent his address, which was read."

In the "Last Public Address by the late Hon. Andrew H. Green, concerning the State Reservation at Niagara," Green summarized four kinds of government parks, briefly recapitulated the movement to set the Niagara Reservation aside and the principles that guided the movement. He also exonerated the officers of the Commission of the State Reservation from implications of wrong-doing.

Almost one year previous to the day (June 13, 1902), with passage in Congress of the River and Harbor Bill, Green had, what must have been the exhilarating satisfaction of drafting federal legislation to protect the Reservation and having it passed on the federal level. One wonders what his emotions might have been toward delivering a speech in the city whose industry the regulation of which he was a principal in achieving.

Much of the high moral tone and deep compassion of the speech is characteristic of Green's personality: "I refer to the uplifting moral effect of the contemplation of those objects which strikingly turn one's thoughts to the great origin of things." The lesson of Niagara Falls is to teach "the inherent right of the people to the free enjoyment of the wonders of nature." Characteristic is Green's deep interest in original scientific research derived from examination of the environment and history at

Niagara Falls. Niagara Falls is "preeminent" for the study of fauna and flora "in their natural habitats." "In the falls themselves, we see nature actively at work in her own workhouse, so to speak, and where, in the storied walls of the great gorge, we find her opened book presenting in extraordinary fashion the record of past ages"—a reference here to the significance of the rock strata in establishing the immense age of the earth, now known as "geological time."

Breaking out of a discourse on the impact of natural phenomena on raising "the thoughts of [the State's] people toward those lofty and sublime conceptions of the Deity" is a series of rhetorical quotes from a variety of sources. This intrusion of rhetorical devices, dropping a discourse of ideas and descending into a list of quotations from the Bible, Emerson, Pope, to illustrate a point, does not appear to match the degree of sophistication characteristic of Mr. Green.

The first is very peculiar: "Then the early French explorers first beheld Niagara, they found the Indian worshipping his Manitou in the falls, rendering to Him peace offerings of tobacco every time they passed, and the sacrifice of two human lives every year."

Although the next three paragraphs offer citations for the information, or quotes, contained in them, there is no reference cited for this information on the behavior of Indians at the falls. The definiteness of the two facts, the offering of tobacco "every time they passed" and the "sacrifice of two human lives every year," is striking, and the implication is that the author of this speech had an historic reference from which these facts were taken.

Another discordant quote derived from the "eminent scientist," William Thompson, Baron Kelvin of Largs (Lord Kelvin) of the 14th Annual Report for 1897, who actively sought to destroy the cataracts and divert all, that is, the entirety of its waters for the generation of electricity. He had wished he could live to see the day when Niagara was dry and hoped his grandchildren would never see the water run free at Niagara. As discussed above, Kelvin had presided over an international organization to devise the technology for generating electric power at Niagara sponsored by the Niagara Falls Power Company. That Green would quote Kelvin in any speech celebrating the great social achievement of the Reservation at Niagara seems rather incredible.

Actually, this lapsing into quotations from well publicized authors is a characteristic of the address given on the death of Thomas Welch, printed separately in this annual report, by Peter A. Porter, Jr. In it he quoted Euripides and an uncredited philosopher who apparently "cried out: 'Mark the difference between intimacy and friendship.'" He quoted Darwin, Ralph Waldo Emerson, Robert Hall, Robert Browning, the Bible, various quotes without attribution and St. Paul (Robert Hall was an "English Baptist minister [1764–1831] and considered the foremost pulpit orator of his time." He was well known for "Modern Infidelity Considered with Respect to Its Influence on Society" published in 1800, according to Bridgwater and Kurtz, 1963).

This technique of quoting authors in speech writing appears to have been to impress upon the audience the wide scope of the reading, if not the education, of the speaker. Reliance upon the ideas of published writers, perhaps, may indicate some uncertainty regarding the quality of one's own. For example, Mr. Porter used this device, thinking that the words "of the great minds of the past, as to what a true friend, and as to what true friendship, is and should be" would be superior to his own. Such uncertainty does not appear to have been a characteristic of Mr. Green.

At a juncture between the two great subjects, or themes, of the speech said to have been written by Green, between the four principles along which public lands (parks) were set aside, and the preservation of Niagara, the speech addressed a third issue: utilization versus waste: "... let me emphasize the fact that as against the utilitarian theory of the Governor to whom I have alluded, this Reservation was created distinctly and purely from the higher aesthetic considerations of which I have spoken." The Governor was Governor Cornell who had opposed the bills to establish the Reservation, stating that the views were a luxury "and why should not the public pay to see them?" (Cornell, quoted by Green, 20 Ann Rep Comm, 1904).

Then the speech asserted an intriguing idea:

"It [the Niagara Reservation] was not established under any subterfuge of utilitarian purpose."

It is interesting to speculate what subterfuge could have been enacted by the establishment of the Reservation at Niagara. If there was a subterfuge, it apparently had to do with commercial use of the river's waters.

"The title of the law which Governor Cleveland signed April 30, 1883, declared that the act was "to authorize the selection, location, and appropriation of certain lands in the village of Niagara Falls for a State Reservation and *to preserve the scenery of the Falls of Niagara* [the emphasis in the speech]. I want to call your attention particularly to this fact, that there was no equivocation in declaring the purely aesthetic purpose of this Reservation, for therein lies its great and distinguishing value as a precedent" (Green in 20 Ann Rep Comm, 1904). This latter seems characteristic of Green's speech.

The Niagara Reservation, in short, was not established to raise property values, enhance the income of railroad and trolley lines, support the tourist industry, or facilitate industrial and utility development along the Niagara River.

"This brings us to the story of how the Reservation was made." By this time, the events which appear to have led up to the creation of the Reservation had been the subject of many retellings. Retellings would continue into the twentieth century, and every time the "story" was told, a little more information would be divulged, as will be discussed below.

One might note that Thomas Welch had in the month the speech was delivered, been elected the next Secretary of the Commission, replacing the historian Peter A. Porter, Jr. The previous year, in 1902, Thomas Welch had himself ventured into historical literature by having an history of the "movement to free Niagara" published by the Buffalo Historical Society "How Niagara was made free, the passage of the Niagara Reservation act in 1885," (Vol. 5:325-329).

Again, almost literally, the curious statements of the history in the 19th Annual Report recur. [In 1869:] "there did not exist, to my knowledge, a single state or national reservation created for the sole purpose of scenic preservation. There was no precedent to which to appeal, and the salvation of Niagara had to be worked out as a new proposition and upon entirely new principles so far as legislation was concerned."

Again, it is curious Green would have forgotten Yosemite, just as it had been in the nineteenth annual report earlier, contradicting the fact that Green had discussed that reserve as early as the seventh annual report, in 1891. Green was a long-standing colleague and sometime friend of Frederick Law Olmsted, who was head of the Yosemite commission before returning to New York City and continuing his work on the New York City parks, particularly Central Park.

Then again, could the members of the American Park and Outdoor Art Association have been ignorant of Yosemite, and would they have accepted Green's lack of knowledge of it and its role in the American park movement? This organization was the first step in creating a national organization for landscape architecture, a discipline that was struggling for legitimacy in academic and professional life in the United States. This association was founded in 1897 in Louisville, Kentucky, apparently through Charles Eliot who suggested "organizing, not a professional, but a general association, to be made up of all who desire the

advancement of landscape art ... in our country. In such a general association, amateurs, ... foresters, gardeners, and anybody interested might
become members" (Charles Eliot, quoted by Newton, 1971). "The Association engaged the interest of numerous landscape architects, including
[Warren H.] Manning" an apprentice of Frederick Law Olmsted, having
worked eight years in Olmsted's firm, and who had just opened his own
office in Boston in 1896-97 (Newton, 1971).

Newton indicated that the American Park and Outdoor Art Association did not last for many years, but it endured at least seven
years—long enough to have a meeting at Niagara Falls in 1903 where
Frederick Law Olmsted, Sr., Calvert and Downing Vaux, Samuel Parsons, Jr., had played so large a role. Perhaps it was to commemorate
these men that the 1903 meeting was held there. Mr. Olmsted, senior,
was gravely ill at the time, and would die a month later, in August. It is
assumed that members of this organization may have known F. L.
Olmsted, Sr. well, including his important role at Yosemite.

Another peculiarity of this speech is the personal intimacy of
Green's direct references to Thomas Welch as Superintendent of the Reservation, and the large role he is said to have played in the "free Niagara"
movement. Reference to Welch's participation concerns the evolving
legislative movement. Much reference is made in the speech to details,
such as the "last minute" character of Governor Hill's signature on the
bill, revealed in Welch's just published history of the "passage of the Niagara Reservation Act in 1885" (Welch, 1902).

The speech indicated that the Niagara movement "had no better
friend or more able helper than the present Superintendent of the Reservation, the Hon. Thomas V. Welch, who was then a member of the Legislature from Niagara Falls, and who, in the most public spirited and disinterested manner, bent every energy to accomplish the end in view."

Green, in his speech, wrote that "You have asked me to say
something about the administration of the Reservation, but time will not
permit me to enter upon the extensive details of this interesting branch of
the subject. The Reservation speaks for itself." Green asserted that it was
administered according to Olmsted and Vaux's plan, "to restore the environment of Niagara as nearly as possible to its natural aspect ...."

Celebratory reference is again made to the Commission as discussed in the nineteenth annual report of the previous year. Of the previous three Commissions, it is the fourth which receives special attention.
"The greatest work of improvement along these lines [of preservation
and restoration and the Olmsted-Vaux plan of 1887] has been accomplished during the past five years under the Commissioners who have

just been re-appointed by Governor Odell." Connection is also made between the Commission honoring themselves in the previous year and their reappointment by the Governor.

It may be that the point of the entire speech, doubtfully written in its entirety by Green, is the following endorsement:

> "From the very inception of the Niagara movement, down through the campaign culminating in the law of 1885 and through the administration of four boards of Commissioners, no touch of self-interest or corruption has tarnished its fair record."

A final comment — The diligent reader of this book and its companion volume "Botanical Heritage of Islands at the Brink o9f Niagara Falls" (Eckel, 2013) has doubtless noticed that despite (1) the title of this book "Ecological Restoration ...", and (2) despite the original legislation's intention for the Commission to at least approximate the equivalent of modern ecological restoration at the Falls, there was essentially nothing of the sort done in the first 20 years of the Reservation's existence. This policy continues to the present day.

A PERSPECTIVE ON THE LEGACY OF THE EARLY YEARS

The responsibility of the parks administrations of New York State is, when stated as a principle of government, to preserve the integrity of the unique resources of each park area. This is achieved by faithfully pursuing the agenda put forth in the legislation which initiated incorporation of specific park areas into the State Government for the benefit of the people of New York State and their descendants.

The primary mission stated as an administrative aim of the park system of New York State is to reserve areas of special natural history, historic significance and to preserve sufficient open space for the purposes of passive and active recreational and educational activities. This mission reflects the complexity of the avowed mission of the New York State Department of Parks & etc. and the complexity of resources inherent in the different parks themselves.

The difference between the two agendas rests in an attempt to prevent changing government administrations from altering the master plans of different parks to suit new and ephemeral agendas based on temporary social problems. For example, building tennis courts in the middle of a biologically important woodland because it is decided the cities are overcrowded. The park was established to protect the woodland. New land should be acquired if a new social agenda is to be pursued without violating the integrity of the legislation that created administration policy toward the woodland.

It follows from the above that local and national parks administrations are beholden to the public that they serve to uphold the laws and administer the lands designated as parks under their custodianship as the natural and cultural heritage of the people they serve or for the health and well-being of citizens as designated in the initiating legislation that removed respective private land into the public domain.

The relationship existing between the New York State Office of Parks, Recreation and Historic Preservation and local governments in the Niagara Region park areas along the Niagara River of which I have some knowledge is probably the most complex of any State Park facility in the State Park system. The governments and government agencies involved are as follows:

Federal Government — The United States Department of the Interior offers some oversight over maintenance of the original custodial mandate of the Niagara Reservation at Niagara Falls, since this is a designation

National Landmark related to the legacy of Frederick Law Olmsted. The relationship between this agency and State Parks is distant and perhaps seldom activated unless intense public interest in the integrity of the site is evident. The only leverage this agency appears to have is to grant or withdraw grants for maintenance of the landmark integrity, or withdrawal of the designation itself.

Interconnecting State Agencies —
   1. *Power Authority of the State of New York.* This agency and that of the NYS Dept. Parks RHP originated under the authority of a single individual, the chairman of both agencies, Robert Moses. Consequently, at Niagara Falls parks issues and power generation issues were inextricably connected and still are, only the interconnections are not as obvious as they were, and local interactions between the two agencies are not necessarily known to the Albany offices of the Parks Department.
   2. *New York State Department of Tourism.* This agency interacts with local tourist agencies, both private and public, using the natural phenomenon of Niagara Falls and its associated natural landscape for the economic benefit of the City of Niagara Falls and the economy of New York State as a whole. The fact that there is a New York State Park in trusteeship of these areas is not an obvious consideration of these agencies and not a conspicuous part of their drive for commercialization of the area. Such commercialization is in conflict with the legally mandated State Parks agenda.
   3. *New York State Department of Transportation.* The construction and maintenance and legal appropriateness of policy proposals for State roadways and transportation facilities such as bridges serving and physically adjacent to the State Parks along the Niagara River frequently interacts with the State Parks due to the impact of landscape modification (deterioration), necessity of parking facilities and the channeling of numbers of vehicularized populations into park areas with inadequate space, not to mention mowing and tree-clearing practices near areas of great biological interest are some of the necessary interactions between this agency and Parks.
   4. *Bridge Commissions.* There are at least two and possibly three bridge commissions controlling four international bridges crossing into Canada from Niagara Falls and Lewiston, New York. Extensive landscaped areas serving these bridges as car and toll plazas and railroad facilities are currently in expansion. Three of the four bridges involve land bordering on within a New York State Park. At least three out of the four

bridges are today currently in expansion, especially that affecting the Niagara Reservation.

Local government — The City of Niagara Falls government has been in an economic crisis for at least the past decade, due to loss of several of its hydro-electric dependent industries and has been seeking to base a new economy exploiting the natural area of the cataracts and the Niagara River—all in State mandates. It is critical to the success of this new economic plan to have economic access to the park areas and has sought to form coalitions with the Power Authority, State Parks and the State Highway Department to use State land to reorient the City toward economic exploitation of the river's edge.

Canadian Government — The two cities of Niagara Falls (New York and Canada) have had an economic partnership in economically exploiting the Niagara River with a focus on the Cataract Prospect that stretches over two centuries. The river edge on both sides of the river has been controlled by government agencies with nearly identical legal mandates to preserve the integrity of the natural environment for over one century. Originally the two parks agencies were to be united in the aim to preserve a natural resource that happens to be under mutual ownership.

Although as far as I know, there is no interaction between the Ontario and New York Park agency to attain the common goal of preservation. There is active interaction between the two city governments, and a member of the staff of the local New York State Parks administration sits on several joint bodies discussing such things as the illumination of the Falls and other issues concerning the use of the river edge and the local city governments. Agents of both the Power Authority of the State of New York and Ontario Hydro also sit together on some of these committees.

The primary change I would like to see in these relationships is that based on the overwhelming natural and economic importance of the Niagara Reservation and other State Parks on the Niagara River associated with the Niagara Gorge to the people and government of New York State, and the nearly overwhelming multitude of administrations with interlocking interests only a few of which have been mentioned, especially on the local level, that a special representative from the Albany office be appointed to explore all issues. It is my opinion that an official must be appointed to negotiate interactions between administrations with a mandate to ensure that the original and fundamental interests of the State

Parks in place be secured. This individual should be objective and separate from the intense pressure from local interests and State agency agendas other than park agendas. This individual would represent the Albany office but would not be under the influence of other State agencies in Albany.

I am unfamiliar with the entire State Park system, but I would venture to say the following:

1. The Niagara Reservation is only second to Jones Beach State Park in number of visitors per year. It is of enormous cultural (industrial and park history), natural history, scenic and economic benefit to both New York and Ontario. It is deeply fragmented: the entire Niagara River between Buckhorn Island State Park and Artpark need unity under one park agency in order to ensure the integrity and inclusion of all the resources under one administration. It is inefficient and expensive and only the most precious points of interest have been made parks: Buckhorn, Reservation, Whirlpool, Devil's Hole and Artpark—all of which should be united into one entity. The fact that the Power Authority owns land may not be a great impediment since there is evidence that State Park money was used to shunt river property into the Authority during the early days of the Authority development when the Niagara Reservation briefly owned the lower slopes of the Niagara gorge. Certain elements of the inefficiencies of Artpark would probably be alleviated by incorporation into a broader mandate, areas of significant biological interest being located on its grounds. Its cultural mission can be distributed throughout the river system rather than being isolated at Lewiston, and improved interactions with the culturally impoverished City of Niagara Falls might improve the State's ability to benefit the citizenship. Also the Authority was considering making land available to the local government, so negotiations must be possible. But even as things stand, even inefficiently designed as it is, the chain of small Niagara River parks can be demonstrated statistically to be of immense significance to the people of New York State.

2. Otherwise, the parks, such as Letchworth and others established in the early part of the century with a superior natural phenomenon at their core must be the nucleus of special administrative interest, for these areas were considered the best resources of their kind in the State when they were established—and they probably still are, even though the State has been explored further since then. The people of the State are irreplaceably losing areas of natural significance set apart in their interests through degradation.

3. Jones Beach probably is a superior example of a park facility having a major beneficial impact on a nearby congested sociologically extremely complex urban area and probably has had an inestimably positive impact on the well-being of that population. It is an outstanding example of a park created in the post war part of this century dedicated to offering a significant recreational outlet for a major city.

A serious management challenge of the state park system is to philosophically resist temporary or administratively-colored ideas of government's sole role as an economic tool as opposed to fundamental principles of stewardship. The administration must maintain alternative concepts of value other than economic wealth on which it must rationalize its priorities. Many of these values are abundantly supported in the literature generated by early parks advocates, and many rhetorical and argumentative positions of great validity may be sought there, such as in the legislative documents submitted to the New York State Legislature by the Commissioners of the State Reservation at Niagara Falls, the literature supporting the acquisition of Letchworth State Park and Alleghany State Park.

Another management challenge is to resist pressures of special interest groups to alter the goals and objectives under which the parks were each established and to substitute new ones. Since park mandates are often established by principle and not expediency, adaptations of the State Parks to contemporary pressures must be met by augmentation, not alteration, by cooperation and collaboration, surrounding ideas and solutions that will mutually enhance public and special interest demands on any one park and the administrative mandates given in trust to the State government.

The third management challenge is greater administrative efficiency in specifying park mandates and goals and the creation of standards or criteria by which mandates and goals can be evaluated as to their achievement or lack of achievement. A very specific challenge is to inject biological expertise into the entire administrative system. The parks hold some of the most precious of the State's natural resources, yet in some cases the sole administrative tool used to maintain the integrity of these resources is a lawn mower. The administrative staff throughout the system with some biological expertise and sensitivity based on a knowledge of natural history is and has been deeply inadequate. Certain educational standards of staff and greater use of professional biological consultants must be implemented. Management plans must be reviewed for their contribution to the biological integrity and maximization of the biological diversity of lands in State control if the Park agency is not to fail in its mandate to preserve the State's special resources.

Recently the DeVeaux School area at Whirlpool State Park just north of Niagara Falls along the Niagara gorge was developed with the construction of a Park administrative building with associated outbuildings for park workers. The Park administration seeded a nearby large field (several acres) with a "wildflower mix" of prairie flora. The field was indeed visually striking but totally inappropriate. A legacy of that decision is the introduction of Cup-plant, now an invasive weed infesting the Whirlpool Park woodland.

For administrative suggestions regarding State Parks along the Niagara River, see remarks made above. The most serious threats to the integrity of Buckhorn Island State Park, the Niagara Reservation, Whirlpool State Park, Devil's Hole and Artpark, although I realize that Artpark is on property belonging to the Power Authority, is their fragmentation. All of these parks celebrate in some way the spectacular geological phenomena of the Niagara River: the broad upper reaches of the river, the cataracts and the seven mile limestone gorge. All the parks are located exactly on landscape foci of the most intense biological interest, channeling a great population of visitors directly to these fragile centers of botanical interest. All of the parks would better be united and a management plan integrating these shrinking centers of scientific interest devised so as to unite biologically all three or four.

With such intense demand for economic enhancement of local economies and the dense congestion of transportation routes with critical points located in park areas on both sides of the border, and with increasing demands made for reducing the hydrological resources by two great generating facilities, there must be a new assessment of park priorities and values, and an official appointed to ensure that park agendas already mandated are actually enhanced by the activities planned for the area. Currently the policies and plans to accommodate local pressures are resulting in a deterioration of park resources both physical and scenic. This is totally unnecessary. Someone needs to study the legal empowerment the parks have in the Niagara region and utilize some of this legal and administrative strength in pursuing park objectives in negotiations with various agencies. This is an opportunity to resolve many centuries-old Niagara park problems.

It appears as though the State's current abundant financial resources dedicated to the Niagara parks are not efficiently used to the parks' benefit. Included in the designation 'financial resources' are the monies committed by other agencies to administer areas marginal to and related to the parks, such as the Department of Transportation suppressing the vegetation all along the river up from Goat Island along the Robert Moses Parkway which, with an adequate biological management plan, could be managed to appear as an extension of the important wooded and biologically significant areas that are the core of the State's Niagara River parks—for example, such a biological, riverside corridor could thematically and scenically unite Buckhorn Island State Park and the Niagara Reservation.

The Power Authority "owns" or otherwise controls important land corridors such as the land along the river just mentioned and the lower bank of the Niagara River along the gorge. I believe that State Parks has some option to put pressure on this agency to maintain these areas in accordance with Park objectives, and the Park agency should have greater say in management and development decisions. An overall biological and cultural management plan would unite overlapping mandates enhancing much of the areas along the Niagara River currently being denuded.

A great deal of money is granted the Niagara area by the New York State Department of Tourism. This agency could be encouraged to participate in Park objectives by working out a way of inserting an ecotourist agenda into their present commercializing and development agenda to augment the commitments of both agencies which presently mount contradictory and competing emphases, with the object that the wilderness and natural resources of the parks have unique tourist value. Recently, millions of dollars are being pumped into development of the Niagara Park System by the New York Power Authority, following agreements associated with relicensing. There is no significant governmental oversight because the monies involved are not from taxes, and the Authority is semi-autonomous.

A substantial increase in financial resources at Niagara, without a clear understanding of the park's mandate and agenda, nor the will to work for the park's interest and augmentation of benefits to the parks, can lead to disastrous decisions, such as to erect additional buildings on park space already congested with administrative structures, to open parking areas without consideration of the enhanced influx of people into already overstressed ecosystems, the preservation of which forms the core of the Olmsted-legacy on Goat Island, which it is the federal government's mandate to oversee. There might be the possibility that an increase in financial resources might free the park administration to remove buildings and other structures, to transfer agents and facilities currently located within the Reservation elsewhere, and generally correct the deteriorated condition of the park, freeing it from its urbanized appearance. Additional money might be used to pay for initially expensive "start-ups" for new administrative policies more in keeping with the area's special interest to the people of New York State.

For example, the mission of the staff of the Schoellkopf Geological Museum should be located on Goat Island, rather than on Power Authority property, possibly in the current maintenance facility or one of the three restaurant concessions at least one of which must be redundant (preferably the larger one). Other than the parking areas, probably it is the maintenance facility that is the least appropriate of all the structures currently on Goat Island in the Niagara Reservation.

Perhaps the most important way to strengthen the public's interest in the State Parks is to ensure that they appear to be well managed. The most serious visual indication that a park is not maintained is due to the inability to adequately maintain natural areas. So much attention is paid to seeding, weeding and mowing lawns that the boundaries of adjacent natural areas appear unkempt because the weeds brought in by lawns and garden activity flourish in the natural areas. Significant reduction in high-maintenance areas in parks with prominent natural areas, or buffers between maintained and natural areas can diminish the unkempt appearance of natural areas. One of the worst processes that can happen when there is a management plan for lawns and none for natural areas may be seen on Goat Island. The natural areas set aside and protected by Olmsted and the focus of the park's mandate are reduced every year by attempts to reduce or control the weedy margins of the lawns, by expanding the lawns into the wooded areas, a process that generates more weeds and more lawn until there is nothing but a weed-infested natural area that the unknowing public cries out to be "maintained." Lawns and garden beds contradict the mandate for the Park, generate unnecessary expense and public criticism of the quality of the Park administration's trusteeship of their land.

APPENDIX 1. THE BRIDGES

Prior to 1848, there were no bridges across the Niagara River anywhere along its approximately 35 mile length.

Judge Samuel DeVeaux, who endowed DeVeaux College at Niagara Falls, New York, upon his death in the 1850's, was, among other business, "interested in facilitating traffic between the Canadian and American sides of the Niagara River to bring growth and prosperity to Niagara Falls rather than to Buffalo. In this capacity he was one of the organizers of the Suspension Bridge Company in 1847 which proceeded to build the Niagara Suspension Bridge across the river joining the Canadian shore with the new community which rose on the American side: Suspension Bridge" (Loker, 1963). The Suspension Bridge Company was organized August 21, 1847, and "included General Charles B. Stuart, former Governor Washington Hunt [of Lockport], and W. O. Buchanan, owner of the Maid of the Mist (whose landing at the time had its entrance near this new bridge site)" (Loker, 1963).

Suspension Bridge — This is the "Suspension Bridge near Niagara Falls" lithograph from Currier and Ives, New York, 1857. The International Suspension Bridge is the first working railroad suspension bridge, completed in 1855 by A. J. Roebling.

The rivalry with Buffalo, indicated by Loker above, probably is a result of the loss of commercial importance of the village of Niagara Falls with the by-passing of the Ontario-Erie Lake shipping routes requiring the portage of goods from Lewiston to Black Rock, New York, and vice versa, with the opening of the Erie Canal in 1825. Buffalo had by then become an important port on the Great Lakes shipping lanes.

It is interesting that W. O. Buchanan, "owner of the Maid of the Mist" was a member of the board of the Niagara Suspension Bridge Company. According to Seibel (1990), in "April 1846 the Niagara Falls Ferry Association received a charter from the State of New York, authorizing it to operate a steamboat ferry across the Niagara River. It is assumed that permission was also received from the Canadian authorities. In May 1846, the clumsy barge-like vessel with two smoke stacks was launched at the new ferry dock located on the American side about 0.8 km (half of a mile) upstream from the present Railway Arch Bridge ... The new ferry, christened "Maid of the Mist," was expected to become a vital link in a proposed new stage coach route from New York City to Toronto" which never materialized (Seibel, 1990). "In 1848 the ferry business lost the majority of its business to Ellet's Suspension Bridge which had just been built across the gorge at the head of the Whirlpool Rapids. The Maid of the Mist then began carrying sightseeing tourists upriver to the Horseshoe Falls." Actually, it appears as though the owner of the steam ferry had expanded his business by diversifying his transportation services, and subsequently used his boats as recreational alternatives to the more comfortable crossing above, if not committing his ferries to the short but thrilling excursion into the spray zones of the American and Horseshoe Falls which we now know as the "Maid of the Mist Ride."

Buck (1898) indicated that two companies were chartered in 1846 to build a bridge across the Niagara Gorge. The Canadian interests were represented by the Niagara Falls Suspension Bridge Company, and the American by the International Bridge Company. In Greenhill (1984) referred to "both the American and Canadian Niagara bridge companies" when discussing the first bridge across the Niagara River, the Niagara Suspension Bridge, which was a temporary structure built in 1848. It was a service bridge, "a fragile structure of wooden towers, and was designed as a temporary span to be used in the construction of a permanent railway and road bridge" (Greenhill, 1984). At the time of the construction of this bridge, the only railroads in its vicinity was the Erie and Ontario railroad (horse-drawn) on the Canadian side, and the "short roads of the Buffalo & Niagara Falls and the Lockport & Niagara Falls" railroads (Greenhill, 1984).

This bridge was replaced by the more permanent Niagara Railway Suspension Bridge, built in 1855 by John Augustus Roebling, designer of the Brooklyn Bridge. This bridge lasted until 1896 when, for structural reasons, the it had to be replaced. It was a railroad bridge with a carriage and footway deck below that of the rails. In 1853, before the bridge was finished, "the upper [railroad] deck was leased to the Great Western Railway" (Greenhill, 1984) with rails to accommodate the gauge differences of the New York Central, the Great Western and the Canandaigua & Niagara Falls Railroad, which implies that these companies all had access to this bridge. An engraving of 1859 of this bridge was titled "The Great International Railway Suspension Bridge. Over the Niagara River connecting the United States & Canada, the New York Central and Great Western Railways" (Greenhill, 1984).

Some 46 years later, it was Roebling's descendant's company, the John A. Roebling's Sons Company of New York City, who repaired the two remaining suspension bridges joining the Three Sisters Islands on May 9, 1901, for which, on May 16, that company was disbursed $2,800.00 (Welch, in 18 Ann Rep Comm 1902, for 1901).

In the period between 1888 and 1893 when the Commissioners of the Niagara Reservation pursued the construction of a bridge north, that is, downstream of the Whirlpool there were already three bridges spanning the Niagara Gorge. Green's bridge would have been a fourth. The proposed bridge appears to have been intended as a match for the Upper Suspension Bridge (Niagara Falls and Clifton Bridge) built within 300 feet of the cataracts in 1869 (Greenhill, 1984), and both would have been light carriage and pedestrian bridges, rather than the more substantial railway and freight bridges. The Upper Suspension Bridge was constructed by a Canadian engineer, Samuel Keefer.

As a matter of fact, a fourth bridge was proposed by someone much later, in 1909, when a petition was made to the New York State Legislature to erect a bridge downstream from the Railway Suspension Bridge and the Cantilever Bridge, and downstream from the Whirlpool north of the property of DeVeaux College, perhaps near the Devil's Hole. The petition was called the Lower Bridge Bill "for the construction and maintenance of a bridge and the necessary approaches for the passage of pedestrians, on the right side of the Niagara river at some point in lot No. 31 of the New York State Mile Reservation to some point in Canada on the left side of said river" (26 Ann Rep Comm, 1910). The Commissioners of the State Reservation opposed the bill, the bridge appearing "to have no justification in a general commercial need of the locality or of the citizens. Its purpose is to deface the scenery of one of the most wonderful parts of the Niagara river, that an opportunity may be given for collecting petty tolls of visitors to the Falls ..." (26 Ann Rep Comm, 1910). This bridge was to have been built such that it did not have the capacity to carry vehicles, and, as such it appears to have been destined to be another concession for tourists, like the ferry rides of the Maid of the Mist boats. Governor Charles E. Hughes vetoed the bill.

The Upper Suspension Bridge was only 10 feet wide and could accommodate a single carriage. It was primarily a tourist, rather than a freight bridge— a "carriage and foot bridge" commanding a spectacular view of the falls. There had been initial opposition to construction of this bridge, some from the Niagara Falls Suspension Bridge Company, the Canadian bridge company that had backed the first railroad suspension bridge (see above), but around 1867–1868 the charter was obtained and the bridge, within 300 feet of the falls, was erected (Morden & Leslie, 1938). Note that the engineer, Samuel Keefer, in 1869 and in Brockville, Ontario, submitted a report to two Canadian bridge companies, the Niagara Falls Suspension Bridge company, and the Clifton Suspension Bridge Company, which was probably the Canadian factor in ownership of the Upper Suspension Bridge (Keefer, cited by Greenhill, 1984, note 29). The Upper Suspension Bridge was also called the Niagara Falls & Clifton Suspension Bridge, perhaps reflecting the names of the two bridge companies (note that assigning nationalities to the bridge companies named is only tentative because of ambiguities in the literature reviewed).

It is with the opening of this span that "competition for attracting visitors to the American side increased," with the Porters in particular adding "buildings, walkways and bridges throughout [Prospect Park]" before selling the park to the Prospect Park Company in 1872 (Scott & Scott, 1983, p. 34). The Porters also built bridges to the Three Sister Islands, and "Goat Island remained the most popular attraction in the 1860's (Scott & Scott, 1983). As a testament to the Porter's earlier commitment to tourism, they had in 1854 "built the new three story brick 'Frontier Mart' on Falls Street for six new stores" (Scott & Scott, 1983. p. 34).

In 1888 an attempt to widen the bridge to 17 feet resulted in disaster for "during a violent storm on January 10 [1889], the trussed deck broke away from the cables and fell into the river. A single traveler, a doctor, just succeeded in reaching the American shore before the bridge began to collapse. It was quickly rebuilt, only to be replaced a few years later by the Upper Steel Arch Bridge" (Greenhill, 1984).

Green's proposed bridge and that of the Upper Suspension Bridge would have formed the ends of a circuit around which the carriaged visitor could see the sights of Niagara on both sides of the river.

The third bridge in existence between 1888 and 1893 was the Cantilever Bridge (Canada Southern, or, Michigan Central Cantilever Bridge) built just upriver to Roebling's Niagara Railway Suspension Bridge, consequently commandeering the view up the gorge toward the famous cataracts for the train cars passing over the bridge. The Cantilever Bridge was built in 1883 for the (William) Vanderbilt railroad lines (Commodore Vanderbilt had died in 1877): the New York Central, Canada Southern, Lake Shore & Michigan Southern and the Michigan Central railways (Greenhill, 1984). The Great Western Railway, which had leased the Railway Suspension Bridge in 1853, had merged with the Grand Trunk line, a "formidable competitor [against Vanderbilt's New York and Michigan Central lines] for traffic between Chicago and the Atlantic seaboard" in 1882. The Grand Trunk had built a bridge far upriver, between Black Rock and Erie, Ontario, near Buffalo, New York, so the New York Central had a railroad bridge of their own built upriver of the Niagara Railway Suspension Bridge at Niagara Falls (Greenhill, 1984).

Interestingly enough, as noted elsewhere, the first chairman of the Canadian Commission for the Queen Victoria Niagara Falls Park was Casimir Gzowski, a civil engineer, who had contracted with the Grand Trunk Railway for building a rail line between Toronto and Sarnia (Seibel, 1985). He was "engineer for the building of the International Railway Bridge across the Niagara River between Fort Erie and Buffalo" (Seibel, 1985).

In 1851, there had been a bridge built at the north end of the gorge called the Lewiston-Queenston Suspension Bridge. This bridge was destroyed in 1864 (Greenhill, 1984). "It was not until the end of the century [in 1899], when electric railroads had been built on both sides of the river, that it was decided to rebuild the Lewiston & Queenston Suspension Bridge to connect the American and Canadian lines" (Greenhill, 1984). The year before, 1898, the Falls View, or Upper Steel Arch Bridge had been rebuilt to accommodate "increased traffic and the plan to provide a double-track connection between the electric railroads on both sides of the river" (Greenhill, 1984). The trolley of the Niagara Falls Park and River Railway connected with the Great Gorge Route trolley across the Lewiston-Queenston Bridge of 1899 (Seibel, 1990, p. 194).

The "Belt Line" trolley system was established with the construction of these two bridges, which was "part of the Great Gorge Route" (Seibel, 1985, p. 220).

The building of both bridges corresponded to the installation of a double-track trolley line on the Riverway in the Niagara Reservation, and all were in preparation for the millions expected to attend the Pan-American Exposition in Buffalo, in 1901.

The two bridges were the Upper Steel Arch Bridge (Falls View, 1898) and the Lewiston & Queenston Suspension Bridge (1899).

Caryl Ely was the president of the Niagara Falls and Suspension Bridge Railway Company. This company and "... allied roads and [roads] under the same general management, to wit: The International Traction Company, Buffalo and Niagara Falls Electric Railway, Buffalo and Lockport Railway, the Niagara Falls Park and River Railway Company, Buffalo Railway Company" were all to be able to use the trolley depot in the Reservation by 1901. When these companies were all merged later under the International Railway Company, Mr. Ely was its President (Mizer, 1981). It is likely that the International Traction Company of Philadelphia (Goldman, 1983) was the holding company for the International Railway Company. According to Peter A. Porter, Jr., it was the International Traction Company which "controls largely the trolley lines on both sides of the river at Niagara, and also two bridges across the gorge" (17 Ann Rep Comm, 1901).

As will be discussed at length elsewhere, Mr. Ely was a director and trustee of the Niagara Falls Power Company. Under the complex "corporate veil" of that company was hidden the identical commercial interests of, by 1901, all the trolley lines on both sides of the river and their associated concessions, bazaars, inclined railways, etc., two international bridges, three power companies: the Niagara Falls Power Company, its Canadian affiliate the Canada Niagara Falls Power Company, the powerhouse of the International Railway Company in Queen Victoria Park, and lastly members of the Commission of the State Reservation at Niagara who controlled the majority vote on issues brought before it (a law passed in 1894, directed that "a majority of said commissioners shall constitute a quorum for the transaction of business," 21 Ann Rep Comm, 1905).

APPENDIX 2: THE RAILROADS.

The issue of bridges was secondary to the railroad issue at Niagara Falls. The bridges were, in fact, aids to railroad passage, including the time when local railroad lines operated electric trolleys. The Commission of the State Reservation at Niagara became dominated by the railroad interests of the Niagara Falls Power Company in 1899. Mr. Thomas Welch, however, had been associated with the Reservation as its Superintendent only a few months before becoming one of the offending interests behind the formation of the power company (Adams, 1927). Green was overruled by Commissioners Dow, Kingsford and Raines, who permitted a street surface railway "upon and along that part of the riverway ... between Falls and Niagara Streets," an issue requiring a special legislative act which was granted by the Legislature in the same year as the appointment to the Commission, by the Governor, of Alexander J. Porter, a director of the Niagara Falls Power Company. The concession to the Niagara Falls and Suspension Bridge Railway Company to build on the Reservation was a concession to an agglomeration of railroad lines under one management involving Mr. Caryl Ely, president of the railway company and also a director of the power company (17 Ann Rep Comm, 1901).

Almost form the beginning, by 1886, Mr. Welch had provided a "market survey" of railroad statistics pertaining to rail excursions to the depots in the village or city of Niagara Falls, New York, including the dates, number of cars, number of visitors and points of departure and on what lines. These surveys were included in all the annual reports since the second report of that series and Mr. Welch's first Superintendent's report in that document. Ostensibly, these railroad surveys were to provide evidence of the value and success of the State's initiative in establishing the State Reservation by showing the interest visitors took in this public land. Visitor statistics could also be used to calculate the savings to the public by making Goat Island and associated land free of access based on entrance charges once levied by the Porter family, the former owners.

These surveys were also surveys of the movements of the visitors on which the tourist industry of the associated municipality of Niagara Falls depended. The village infrastructure of tourist shops, hotels and concessions was, to a great extent, organized around the depots or terminals of the railways.

The Porter family of Niagara Falls, New York, had had an interest in developing the tourist, transportation and manufacturing foundation of the settlement beside the cataracts from the period when Indians still held title to the lands, as discussed earlier. Augustus Porter had made some of the initial surveys of the primitive lands along the Niagara River when the frontier of the United States reached as far west as Canandaigua, New York. He and his brother, Peter Buell Porter, and associates had made strategic purchases of land along the river, and were to continue to have a strong interest in real estate acquisitions throughout the nineteenth century. The initial edge the family had, when it controlled or attempted to control the water and overland transport routes on Lake Ontario to Lake Erie and the necessary portage form Lewiston to the north on the Niagara River to a point south above the cataracts where the river became navigable again, was destroyed at the opening of the Erie Canal in 1825, and the Welland Ship Canal, Ontario, November 27, 1829.

The portage businesses along the river in Ontario, which had organized themselves around the Portage Road form Fort Erie to the south, and that paralleled the Niagara River, and particularly at Queenston, had also received a rude shock when the Welland Canal Company constructed a ship canal on the Niagara Peninsula in Ontario i 1829, and made redundant the overland portage to docks at Queenston, opposite Lewiston, New York. This canal tempered the advantage Buffalo, New York, gained by being the western terminus of the Erie Canal. Shipping bound for Lake Ontario from the upper lakes could now bypass the hegemony of the Buffalo to Albany to New York canal route, and serve Canadian and northern New York cities along Lake Ontario and the St. Lawrence River (Brown & Watson, 1982).

The cataracts in the Niagara River were a critical obstruction to water transportation throughout the Great Lakes region and local businessmen in Ontario and New York both benefited, at first, from the bottleneck. However, regional solutions to this transportation dilemma were always sought, and ways were found to bypass this difficulty, making the cataracts a liability to the very people who sought to capitalize on its impedance to transport.

The canals, however, probably had little or no impact on a fledgling tourist economy on either side of the Niagara River. The canals were built primarily to haul freight: agricultural goods from the west travelled east, industrial goods manufactured in the east and in Europe travelled west. It was the local businesses dependent upon freight and its handling that suffered with the opening of the two canals. Visitors, apparently, still preferred overland stage coach transport, regardless of its discomfort, because of the speed.

A decent road existed along the Niagara River in British territory, and a portage was established between Queenston and Chippawa by 1791. A stage coach service was available on this route (Seibel, 1990). There appears to have been no important stage line on the American side, or if there was, it was not used as profitably as that on the Canadian side, ultimately running passengers between points at Buffalo-Fort Erie and Lewiston-Queenston in view of New York State.

The same New York businessmen, such as Cornelius Vanderbilt, who operated the sailing- and steam-ships on the Great Lakes, the Hudson, on trans-oceanic lines and scores of other waterways tended to fuel the railroad expansion in New York State. Peter B. Porter, for example, "was the force behind the Buffalo-Niagara Falls railroad" (Brown & Watson, 1982) of 1836, the first between the two cities. "Both the Buffalo & Niagara Falls and the Lockport & Niagara Falls were incorporated in 1834. The Buffalo & Niagara Falls began operating in 1836, and the Lockport & Niagara Falls the following year" (Greenhill, 1984, note 13). According to a traveler's guide issued in 1834, visitors to Niagara had a variety of ways to get there from the east (the west was still the frontier): by packet boat on the canal, by steamboat of the Niagara River from Buffalo or from Lake Ontario and up the Niagara River to Lewiston, New York, or Queenston, Ontario, and by overland stage, but that there was a "railroad established at Lockport in order to bring visitors to the falls, bypassing Buffalo" (Ingraham, 1834), that is, for travelers on the Erie Canal, on which route Lockport preceded Buffalo from the east.

It would be the railroads that would bring their human cargo on which the tourist economy of Niagara Falls, New York, would begin to depend in the 1830's. Once the railroads were established, overland stages were abandoned by the general public except for local routes. People would continue to ride the rails until the automobile came into general use in the early twentieth century.

Across the river, in Ontario, the "Erie & Ontario Railroad was incorporated in 1835, and began operating in 1839 between Chippawa and Queenston," (Greenhill, 1984, note 12). Here, the railway was in competition with the Welland Canal. "Merchants who had lost business since the Welland Canal opened in 1829 ... applied for a railway charter. Opposition from the Welland Canal Company and Niagara held up approval. Finally in 1835, the Erie and Ontario Rail Road Company got its charter" (Seibel, 1990, p. 163-4). This line extended from Chippawa, on the Chippawa, or Welland, River ten miles north to Queenston. "Steamboats continue the line of travel from both ends of this road, thus furnishing an interesting and speedy conveyance between the Lakes Erie and

Ontario" (Holley, quoted by Seibel, 1990, p. 164. The power to drive the train was that of horses. The rails were of wood with an iron strip attached to their upper surfaces. The Lockport and Niagara Falls Railroad was also horse-drawn, also ran on wooden rails, and was called the "Strap Railroad" because of the iron strip fixed to the wood.

Another example of the railroad initiative that took place in the 1830's and an example of yet another transportation route west in New York State, bypassing the Lake Erie – Lake Ontario connection, was the New York and Erie Railroad, chartered by the New York State Legislature in 1832 as a response to the success of the Erie canal, which opened seven years previous. This would become the Erie Railroad, which finally opened in 1851 connecting "Piermont on the Hudson, about twenty miles north of New York City, to Dunkirk on Lake Erie, with numerous branches," effectively linking "the southern counties of New York State with the Great Lakes" (Mountfield, 1979).

Development in the 1820's — The first major hotels at the cataracts were built on the British-Canadian shore probably just after the War of 1812. By 1819 there was a Forsyth's Hotel, otherwise known as the Niagara Falls Hotel, owned by William Forsyth, the "first entrepreneur in the burgeoning tourist business" (Seibel, 1990). Mr. Forsyth was to set the pattern for developing and monopolizing the attractions at the cataracts that was to be repeated many times for two centuries by local businesses at Niagara.

Forsyth controlled a major local stagecoach line and sought to lure visitors onto the line by the accommodations he provided and the entertainments and attractions he began to develop there. He purchased land adjacent to the Horseshoe Falls and Table Rock. He built a covered stairway in 1818 down the gorge wall at the Horseshoe Falls, reminiscent of the Biddle Stairs and the covered stairs to the American ferry on the Porter's property on the American side. Visitors could descend "to reach the river's edge and from there walk behind the sheet of falling water" (Seibel, 1990) a decade or so before Augustus Porter made it possible to descend to the shoreline at Goat Island and walk behind the sheet of what is now called the Bridal Veil Falls, and perhaps the American Falls as well, the "Cave of the Winds" attraction.

Forsyth ran a stagecoach line from Buffalo and Black Rock on the American side, across to Fort Erie (doubtless by ferry) on the Canadian shore, thence north to his hotel near the cataracts and on to Queenston and Niagara-on-the-Lake (Seibel, 1990). It is probable the capital acquired from the stage helped him develop the Ontario shoreline at Niagara. He established the first ferry at the base of the American falls, also in 1818, the year after he bought Wilson's Tavern. The tourist industry on both sides of the river may be interpreted as an outgrowth of the developing transportation industry. People paid to ride to the cataracts and, naturally, concessions and attractions were developed to encourage them to stay and pay for the services established for their convenience. The longer they could be made to stay, the more money could be made off them. The more that would ride the stages, steamboats, ferries and later railroads, the richer would become the transportation businesses.

Almost from the beginning, in 1820, Forsyth endured the slurs of unspecified people, perhaps rivals, who allegedly distorted the actual fees he charged visitors on his stage routes and in his hotel for board, breakfast and lodging "calculated to injure the reputation of his house" (Forsyth, quoted by Seibel, 1990, p. 66). In Forsyth's future advertisements, "he repeated this theme over and over—that rumors were circulating about his charges, etc., and he wanted to answer these scurrilous comments by quoting his prices or giving his side of whatever the current rumor might be. He always gave the impression that someone was maligning him, and that he was innocent of the charges made against him" (Seibel, 1990).

An interesting specimen of the kind of pressure that could be mounted by one interest in the tourist infrastructure at Niagara against another is the series of aggressive attacks in the 1880's made by Thomas V. Welch, as Superintendent of the State Reservation together with businessmen in the village of Niagara Falls against the New York Central and a commissioned coach or carriage company associated with the railroad, as will be discussed below.

Competition with Forsyth began with John Brown, who built the Ontario House, or Brown's Hotel, near Forsyth's Niagara Falls Hotel. In response, Forsyth rebuilt and improved his hotel, calling it the Pavilion. "For a time there was nothing comparable to the Ontario House and the Pavilion on the American side" (Green, quoted by Seibel, 1990). These great hotels stood high on the embankment overlooking the cataracts in full view of the riverbank owners on the American side who could only gaze across the river at this mark of commercial enterprise and gentility, based as it was on the Portage Road traffic and the stage lines that operated at the back of Forsyth's and the front of Brown's hotels.

Brown proceeded to open his own stage-coach line in competition with Forsyth. "At this period (1821) the passenger carrying business on the Portage Road was so profitable that enterprising citizens of the

States came over to participate in it. In 1822, after a petition by Forsyth and others, the Provincial Legislature prohibited aliens from engaging in such traffic" (Green, 1926, quoted by Seibel 1990, p. 69).

Brown and Forsyth began to undercut one another, Forsyth in particular attempting ways to block Brown's commercial success. Forsyth built a fence on his property at the prospect at the Horseshoe Falls to control access to the scenery. Brown appealed to the Provincial Government which responded by pulling down the fence. Forsyth rebuilt the fence, and the government again destroyed it. Due to an apparent adverse effect the publicity of these events had on his business, Forsyth faked a letter by a "traveler," a "self-serving account concocted by Forsyth to obscure what really happened" (Seibel, 1990) to a journal, the Colonial Advocate, informing the public that the actions of the government were not for the correction of commercial monopolization and degradation of the area, but to provide free public access to the cataract. It is interesting that free access to the public was an issue this early in the century. This access was also the right of the British-Canadian populace because the land used by Forsyth on the river margin was actually public property, part of the Chain Military Reserve, corresponding approximately to the Mile Strip reserve across the river in New York.

The Ontario Government ultimately gave two individuals, Thomas Clark and Samuel Street, "exclusive lease to the Chain at the Horseshoe Falls' and official management of Forsyth's covered stairs. Before they had done so, the lease was granted to a Christopher Boughner, whose farm "adjoined the gorge at the ferry landing" (Seibel, 1990, p. 114). To effectively regain the lease privilege from Boughner, Forsyth proceeded to buy up all of Boughner's debts in the community and threatened to take Boughner's land if Boughner did not either pay up or assign Forsyth the lease. Forsyth succeeded, but the Government again intervened and granted Clark and Street the ferry privilege by 1825 (Seibel, 1990).

Again, Forsyth faked a letter to the Colonial Advocate, by a "traveler," essentially disclaiming that his son burnt down Brown's Hotel, the precursor of John Brown's Ontario House: "... the most ungenerous, unjust, and infamous reports have been circulated, (by some malicious evil minded person, or persons as yet unknown) throwing out insinuations, that he (Forsyth) was privy to the burning of the house of Brown and subsequently destroyed his harness, etc." (Forsyth, quoted by Seibel, 1990). This time, when the fictitious letter was published in the Colonial Advocate, "the editor was not deceived and carried it as an advertisement" (Seibel, 1990).

In 1825, the Porter brothers had issued their "invitation to eastern capitalists and manufacturers" Claiming that "a thousand mills might be erected ... and each supplied with a never-failing water-power ...," the forests surrounding the falls would be a source of "a cheap and abundant supply of fuel for manufacturing purposes." Goat Island, its name made more attractive as Iris (Rainbow) Island, "contains about seventy acres of excellent land, the upper half of which might be covered with machinery, propelled by water-power; and the lower half, situated in the midst of the falls and rapids, where Nature courts the imagination in her most sublime, beautiful and fascinating forms, might be converted into delightful seats for the residence of private gentlemen, or appropriated to hotels and pleasure grounds for the accommodation of the numerous strangers who annually visit this spot" (prospectus published by Adams, 1927). Still, in spite of the hydraulic potential, getting freight into and out of the area above the cataracts posed a problem that made establishing manufactures easier in cities along the Erie canal.

A year later, Thomas Barnett built his famous museum, in 1826, in Ontario on the lower level below the Pavilion and Ontario House hotels. This museum appears to have had its exhibits organized according to the earlier Peale Museum, in New York City, discussed below. The Barnett Museum is not the oldest extant in North America and, miraculously, its exhibits existed intact until relatively recently. These have since been sold and largely disbanded.

On the American shore, the hotelman General Parkhurst Whitney operated the Eagle Tavern, later called the Eagle Hotel. By 1834 Whitney owned the Cataract Hotel, as well, which accommodated 100 permanent guests (Ingraham, 1834). He collaborated with Forsyth and Brown on a number of mutually beneficial activities. In 1818, in addition to building the first bridges out to Goat Island from the mainland, "Forsyth on the Canadian side and Whitney on the American side, built stairways down the steep gorge wall to the top of the talus slopes, and then cut winding pathways along the slopes, leading down to the river's edge. From there guests at Forsyth's and Whitney's hotels were able to board a small row boat, to be rowed across the river," (Seibel, 1990).

Another joint venture between the two proprietors was "staging the first stunt, organized and promoted with the avowed purpose of attracting tourists to Niagara Falls. On August 2, 1827 they advertised that on September 8th 'The Pirate Michigan' will pass the Great Rapids and the Falls with a Cargo of Ferocious Animals.'

The event was a huge success in as much as it attracted thousands of visitors who would not otherwise have come to Niagara Falls"

(Seibel, 1990). Five Lake Erie steamboats brought visitors, stages, wagons, schooners brought eight to ten thousand visitors to witness the terrified behavior and death of the animals on board the doomed ship, released upriver from the brink of the falls.

In anticipation of this event, the Porters permitted a bridge to be built out over the water on the New York flank of the Horseshoe Falls to a point of exposed bedrock (Terrapin Rocks, in 1827, Porter, 1900). The bridge was built to compete with the Table Rock on the Ontario flank, from which visitors could gaze down at the falling waters (Porter, 1900). Ingraham (1834) placed the bridge construction a year later: "this bridge or platform, was erected in 1828, by Gen. Whitney, the proprietor of the Hotel, who was also its projector. It is about 300 feet long, and projects over the Falls about 8 or 10 feet." According to Ingraham, General Whitney and Mr. Gad Pierce built the first bridges to Goat Island, in 1818 (the bridge of the previous year having been swept away). Yet it was the Porters who later claimed to have built the bridges: "promptly with the coming of spring, 1818, the Porter brothers erected a second but a more substantial wooden bridge. They selected a site ... downstream and built it form the mainland to Bath [Island] ... and from that island they built another bridge to Goat Island" (Porter, 1990).

The Porters built the Biddle Stairs two years after the unfortunate "Michigan Pirate" event, in 1829 (Porter, 1900). It was erected in the same year as the stunt performed by Sam Patch, discussed below. Seibel, perhaps unaware of the importance of the financier Nicholas Biddle and his role in the Bank of the United States, indicated that these stairs were built "with money provided by a tourist by that name" (Seibel, 1990, p. 117). Nicholas Biddle had been made president of the Bank of the United States in 1823, after being appointed one of its directors by President Monroe in 1819 (Bridwater & Kurz, 1963).

Ingraham (1834) also referred to the Biddle Stairs "erected at the expense of Nicholas Biddle, Esq." Peter A. Porter, Jr. was later to give two versions of Mr. Biddle's financial relationship to the erection of these stairs, first that the stairs were built at his suggestion, and that "he contributed a part of the expense of their erection" (Porter, 1900). A year after Porter's remarks on Biddle's sharing in the finances, Porter wrote that Biddle, "who is credited with having suggested to the owners of Goat Island this means of accomplishing the descent, and is further said to have offered to share the expense of their construction, although this philanthropic part of his suggestion was declined" (Porter, 17 Ann Rep Comm, 1901).

The first of the physical modifications to the cataract environment, outside of logging, was the blasting for reasons of safety, in 1829, of the caprock ledges that projected out from the top of the gorge walls on the flanks of the Horseshoe Falls, of which Table Rock was the most famous. Sam Patch's famous leap form a platform erected on the west end of Goat Island into the waters below was scheduled to coincide with the dynamiting, and another vessel may have been sent over the falls (Seibel, 1990). Augustus Porter must have participated in Sam Patch's event because Porter owned Goat Island, and Parkhurst Whitney was probably the liaison, staging events in concert with the hotels across the river.

Development in the 1830's —In the thirties, in spite of the cholera epidemic that afflicted the northern United States and Canada in 1832, and the economic depression in the United States of 1837, development of the area by the cataracts intensified and commercial emphasis shifted to the American side. The Porters and their associates brought the railroads to their community in that decade, offering what would ultimately become a superior connection to whatever stage coach transport was available between Niagara Falls and Lockport or Buffalo.

In Ontario in 1832, Forsyth, "disgusted and financially drained by his legal costs ... sold his holdings which, by then, included the Ontario House to The Niagara Falls Company, promoters of 'The City of the Falls,' a group of Canadian investors who were buying up the land on the flats facing the cataracts and organizing it into subdivisions for sale to American buyers for summer cottages. They were to call this 'The City of the Falls.' These investors also owned the Pavilion. In an early guidebook for 1834, the Pavilion, 'once kept by Mr. Forsyth and last year by Mr. Crysler is now under the control of Mr. Atkinson, formerly Curator at Peale's Museum, in New York" (Ingraham, 1834).

Harmanus Crysler operated a hotel, the National Hotel, in Drummondville, Ontario, near the falls. He sold the National in 1833 and began to build the world-famous Clifton House near what would become the north end of the Queen Victoria Niagara Falls Park, and the Ontario end of the Upper Steel Arch, or Honeymoon Bridge. It opened in 1836 (Seibel, 1990).

The Porters appear not to have been interested in the tourist business as much as they were in developing Niagara as a manufacturing center. They did not appear to own the internationally famous hotels that came to be built on the American side of the river near the new railroad depots beginning during the 1830's, although they were closely associated with the individuals who did own them. By the time their land, or what they still own in 1880, came to be purchased by the State for the Niagara Reservation, they had assisted in the development of three hydraulic canals and the industries established along them within the mile or so of river upstream from the brinks of the cataracts. For some reason, they never did develop or exploit the natural resources of Goat Island except on the extreme periphery and then only at specific points of special interest.

In the period around 11825, the Porters "were the power users and power developers, but were opposed to any such uses of this Island. They did develop power and erect mills on the main shore" (Porter, 1900).

It was as early as the thirties, too, that the issue of development, which was considered a "sacrilege," could be detected in the press. This idea of the sanctity of the cataract environment was probably a universal, rather than a local sentiment, and was based on the respect shown the travel books published during the nineteenth century by the cultured, intellectual and leisure class. The Niagara cataracts belonged to the world, and local developers were sensitive to too hasty or too obvious decisions to reduce it to economic gain.

For example, a local paper, during the 1830's remarked that "sensitive travelers have bewailed the sacrilegious erection of huge hotels on a spot that should be sacred to the sublime and beautiful ..." (quoted by Seibel, 1990). Niagara was ready to spring into new stages of development on both sides of the river. As to the Canadian scheme, the City of the Falls, the first guidebook to the falls remarked: "I have room only to *allude* to the "improvements" which are "being made" in this vicinity. I look with a suspicious eye upon every measure calculated to mar the beauty of Nature's works; and I think there has been too much of this done, here, already. But I do hope that all future alterations and additions will really be improvements.

"The 'City of the Falls' is laid out on a splendid scale, and many cottages have been erected. It will be a delightful summer residence. I shall give a full account of it, and an engraved plan of the proposed 'city,' in my larger work" (Ingraham, 1836). "Improvements are in contemplation on the U.S. side, also. But my readers must come and see them for themselves."

Part of the "improvements" on the American side could be seen in a map accompanying a travel guide produced by Parsons (1836), where drawn on his map of Niagara Falls, New York, is a "Proposed Ship Canal" leading from the upper Niagara River at a point just upriver from Goat Island, east of Fort Schlosser on Gill Creek opposite Navy Island. The proposed canal trended north to near the Devil's Hole and ended at Lewiston, New York. Agitations to excite Congress into building this canal from at least this period may be characterized as follows:

"In its commercially historic aspect, there stands out one important project in connection with Niagara Falls which has been broached by its advocated in public and in private, and especially in the halls of Congress for the past three quarters of a century. Although by international treaty, no war vessels are permitted on the upper lakes ... the advocates of a ship canal of a capacity large enough to float our largest vessels, connecting the Niagara river some two or three miles above the Falls with its quiet waters at Lewiston or below, have continued their agitations; and preliminary appropriations, and elaborate surveys—showing three or four routes—have been made by Congress at three different times. The project so far has made but little headway towards a successful consideration" (Porter, 10 Ann Rep Comm, 1894).

In 1855 an American ship canal was built in the upper Great Lakes, bypassing the rapids in the Saint Marys river, the Sault Sainte Marie Canal, between Lakes Huron and Superior. In 1895, the Canadian government built a canal at the same site in Canadian territory, thus insuring the autonomy of their shipping lanes. Between them, these canals are considered "the most important and busiest waterways in the world" (Brigwater & Kurz, 1963). The Canadians took the initiative with the Welland Canal in 1829—it would seem reasonable to have created a ship canal on American territory. Perhaps the existence of the Erie Canal and the confidence of the Federal Government cooled political and economic arbor at such a proposition.

In 1833, steamship lines on Lake Ontario pursued a brisk business, as did the overland traffic on the Portage Road in Ontario such that a local paper observed that "we believe there are few places in America where there is so much travelling" (quoted by Seibel, 1990, p. 81).

Local railroad lines were built on both sides of the river. By 1836, Whitney had the Cataract Hotel built in the village of Niagara Falls, New York, and the Eagle Hotel, both at the terminus of the Niagara Falls to Lockport railroad (Adams, 1927). Another hotel in 1836 was being built, the Rathburn Hotel (Seibel, 1990). A short hydraulic canal had been excavated with a wing dam out into the Niagara River opposite the head of Goat Island. The Biddle Stairs had been built and the Terrapin Bridge and Tower—the tower was constructed in 1833.

In Ontario, The Niagara Falls Company pursued their goal of developing a city on the gorge and river bank by the flank of the Horseshoe Falls. They apparently declared they wished to preserve the cataracts from "vandalism and commercial enterprises that would detract from the natural beauty of the surroundings" (company quoted by Seibel, 1990). Their surveyed lots were not on the elevated terrace overlooking the cataracts where the old hotels and the Portage Road had been situated, but down below, on the old riverbed within what was to become the boundaries of the Queen Victoria Park.

"The scheme was doomed when railroads replaced horse-drawn stagecoaches beginning in the mid 1830's as the means by which tourists came to Niagara Falls" (Seibel, 1990). By 1834 the first advertisements of Niagara Company stock were circulated to newspapers in Buffalo and New York City, Toronto, Hamilton, London and Niagara, but shares were withheld for later sales in Great Britain, Ireland and the West Indies (Seibel, 1990). Presumably, according to Seibel, the promoter's failure was due to loss of the tourist trade to the New York side, and yet he does not cite evidence to support this view, and it may be rather an interpretation of subsequent events. Another interpretation might be given for why the "scheme" failed. If lots were to be sold to local residents for cottages, it would appear that tourists were not part of the equation, otherwise the speculators would have invested in hotels and other services with a tourist agenda. The fact that railroads brought visitors to Niagara Falls, New York, only gave impetus to the development of the hotel and service industry there. Hotels were not summer cottages.

It might be presumed that in fact the promoters of the City of the Falls almost certainly had a hidden motive for their enterprise, and that was industrial development of the falls area. Summer cottages were in harmony with a professed interest in preserving the cataract borders from "vandalism and commercial enterprises that would detract from the natural beauty of the surroundings" (company quoted by Seibel, 1990). In fact, the whole rusticized "city" idea may have been a fiction invented to facilitate permits to purchase the area below the escarpment adjacent to the river's edge, and to promote goodwill associated with their enterprise while intending to industrialize the area. That they failed may have been due to the financial panic of 1837, which affected the economy of Great Britain (Mountfield, 1979) as well as the United States and Canada, not the opening of American railway lines.

Why the City of the Falls scheme may have failed during an economic depression and yet the construction of railways in the area carried forward is due to a curious feature of railway development: "it was really a general economic boom that encouraged railway promotion rather than vice-versa, and the ensuing slump, which was anyway due to wider causes, was to some extent [for railway investors] ameliorated by the economic activity generated by the successful railway promotions: to generalize, promotion took place in a boom, construction often in a slump" (Mountfield, 1979).

A Lieutenant E. T. Coke quoted by Green (1926, in Seibel, 1990) revealed that Mr. Coke had made an unhappy discovery upon conducting an investigation of his own: "the company of speculators intend erecting grist mills, store houses, saw mills and all other kinds of unornamental buildings, entertaining the most sanguine hopes of living to see a very populous city. The die is then cast and the beautiful scenery about the Falls is doomed to be destroyed. Year after year it will become less attractive .... Tis a pity that such ground was not preserved as sacred in perpetuum."

So powerful was the ill will that could be generated at plans to "desecrate" the cataracts by commercial development, that the failure of the whole enterprise could be directed by those who lost their investment at an author who wrote a text whose views were well publicized disparaging the scheme and perhaps driving off potential investors—Sir Richard Bonnycastle. Bonnycastle does not appear to have been an ordinary tourist. He commanded the Royal Engineers in Upper Canada and had visited the falls in 1835. He had appended a letter to a report submitted to a Select Committee of the Legislative Assembly of Upper Canada regarding "the subject of a suspension Bridge over the River Niagara or the construction of a Tunnel under the same" (Bonnycastle, 1836, in Greenhill, 1984).

The Notorious Front — This delightful artist's representation gives some indication of the kind of visitor's experience to be had, at least on the Canadian portion of the area of the Falls in the 1860's. No attribution for this cartoon is given by the source (Kiwanis Club, 1984). Drawing by Duncan Macpherson.

Appendices

Commercial buildings in the 1880's — Appearance before the State took over the area as a Reservation (Dow, 1921).

Upon revisiting Niagara in the 1840's he wrote "I was so disgusted to see the spirit of pelf, that concentration of self, hovering over one of the last great wonders of the world, that I rushed to the Three Horse Railway, and soon forgot all my misery in scrambling for a place ..." (Bonnycastle, 1849, quoted by Seibel, 1990, p. 103). Bonnycastle had also publicized a book in 1842 on Canadian themes (Dow, 1921).

What the speculators had done to Niagara deeply disturbed Sir Bonnycastle: "... The Company of the City of the Falls—a most enlightened body of British subjects, who first disfigured Table Rock, by putting a water mill on it, and now are adding the horror of gin-palaces, with sundry ornamental booths for the sale of juleps and sling, all along the venerable edge of the precipice, so that trees of unequalled beauty on the bank above, trees which grow nowhere else in Canada, are daily falling before the monster of gain.

"It is the greatest wonder of the visible world here below, and should be protected from the rapacity of private greed and not made a Greenwich fair of; where pedlars and thimble-riggs, and barkers, the lowest trulls and the vilest scum of society, congregate to disgust and annoy visitors from all parts of the world, plundering and pestering them without control" (Bonnycastle, 1849, quoted by Seibel, 1990).

Perhaps one of the most offensive elements of the company of developers was their hiding behind the pious front of preservation while promoting an enterprise which would degrade, simply because there was no regulatory element in their proposal to prevent this from happening. Their failure also may be considered to have taught a profound lesson to local developers on both sides of the river—the power of the printed word, and the power of world (European and American) opinion regarding protection of the falls of Niagara. In the future, developers would have to be much more subtle, more organized, and they would have to stage a publication campaign of their own to change world opinion from protection of a resource to its exploitation for the good of society. Science, which before had stood for the great biological diversity of life native to North America and nurtured by the productive environment of the cataracts as presented to the world in the natural history displays of Mr. Barnett's Museum, would later become a propagandistic tool in the hands of engineers. Niagara's natural history museums of the nineteenth century would become museums of technology and the exploitation of Niagara's water power in the twentieth. Natural history information, which formed part of the rhetoric of preservation of the falls in the previous century, would be replaced in the twentieth by technological information in the rhetoric of those who wished to exploit the falls, and diminish its power and biological and geological integrity.

Development in the 1840's — One of the charter members of the Niagara Falls Company, Thomas Clark, was also one of those who applied for a charter to build the Erie and Ontario Rail Road Company, which was granted in 1835 (Seibel, 1990, p. 163-4). Only one section of the railway was constructed by 1839, that between Chippawa and Queenston—far enough to make connections on the Niagara River above and below the falls. This railroad was called by the public the Chippawa-Queenston Railroad or alternatively the Queenston-Chippawa Railroad, but was not the official name of the company.

The primary source of revenue for this railroad was as a passenger service. Freight was carried through the Welland Canal (Seibel, 1990). Getting the cars up the three hundred or so feet of sheer drop along the escarpment would be a major impediment for this railway and any other subsequent transport system established in the area.

Although the Ontario Government may not have financially helped with the City of the Falls enterprise, £5,000 was loaned the financially strapped new railway company and the first railway line in Upper Canada, that is, outside of Quebec, was constructed: a horse-drawn strap railroad (Seibel, 1990). Horses were to continue as the mode of power for the line until the company received permission to change to steam in 1852, the year when it received approval in its charter to extend its line from Queenston north to Niagara-on-the-Lake on the shore of Lake Ontario at the north end of the Niagara River (Greenhill, 1984).

In the 1840's again a curious conflict, in which malice may or may not have been used, was prosecuted regarding the Niagara Falls ferry, which was operated by Canadian authorities. Two groups of American citizens of the village of Niagara Falls wrote conflicting statements to the Canadian authority, one proclaiming that the two ferrymen in charge of the boats were "addicted to habits of intemperance and are from that cause unfit many times, when their services are required to take charge of a boat in crossing the Stream ... We have also been informed and some of us know the facts that due to their indulgence in strong drink they are often unaccommodating and uncivilized to passengers and as we are informed are in the habit of extorting from persons larger sums for ferriage than they are permitted by their agreement with you to exact for their service" (1842 petition quoted by Seibel, 1990). A month later, one of the ferrymen received an endorsement by a different set of American individuals who "believe that any objections or complaints brought up against him, to be of an obvious and malicious nature, calculated to injure him in the eyes of the public as well as those by whom he is employed." The second ferryman received an endorsement from a group of Canadian individuals who were "believing any complaints got up against him or the Ferry are of a malicious tendency calculated to injure him" (petition, quoted by Seibel, 1990).

Nearly two decades after the creation of the Erie and Ontario Railway, in 1852, the Briton Thomas Brassey and associates "contracted to build the Grand Trunk Railway of Canada. It was to run 539 miles— at that time the longest railway in the world—from Quebec to Toronto along the valley of the St. Lawrence, crossing the river by the Victoria Bridge at Montreal," (Mountfield, 1979). Its purpose was "to provide reliable, year-round transportation on eastern Canada's main transportation route. ... The Grand Trunk was eventually to provide direct communication between Riviers du Loup (east of Quebec) and Detroit. . The line would encourage settlement, it would help in fostering a sense of national unity among the Canadian provinces, and it would stop the drain of Canadian goods southward to New York, via US railways, which was impoverishing Canadian ports" (Mountfield, 1979).

An official of the Great Western Railway, which in 1882 would amalgamate with the Grand Trunk Railway, unsuccessfully attempted to convince his directors to purchase the Erie and Ontario Railway at Niagara in 1854, anticipating that it would be a lucrative short run catering to visitors taking the Lake Ontario steamers dropping off to see the falls when the rail line was extended to Niagara-on-the-Lake (Seibel, 1990, p. 105). Around 1857, the Erie and Ontario Rail Road Company, perhaps after some corporate reorganization, changed its name to the Fort Erie Railroad Company in anticipation of building a line from Chippawa south to Fort Erie, Ontario. The line apparently went bankrupt and was purchased after 1862, forming a company called the Erie and Niagara, which prosecuted the line to Fort Erie. In spite of another attempt by the Great Western Railway to purchase it, the Canada Southern Railway took over the Erie and Niagara in 1878, two years after the Canada Southern was taken over by Commodore and his son William Vanderbilt of the New York Central and Hudson Railroad (Greenhill, 1984). The Michigan Central bought the Canada Southern Railway in 1904, and "in 1929, the New York Central leased the line from the Michigan Central (Seibel, 1990). These were business details, however, for Commodore Vanderbilt had acquired control of the Michigan Central as early as 1876 (Greenhill, 1984). Control of the early railroad along the Niagara River, than, had been taken over by directors of the New York Central since 1878.

In New York, the Lockport and Niagara Falls Railroad, which had been incorporated in 1834 and operational by 1837, "was taken over in 1850 by a new company, the Rochester, Lockport & Niagara Falls, soon to be part of the New York Central" (Greenhill, 1984). Apparently, the original route was abandoned sometime after the New York Central took over this company. It appears, however, as though Porter's railroad, the Buffalo and Niagara Falls, managed to persist in some form. In a map dated 1836 (a copy printed in Ingraham, 1834, reprinted by Adams, 1927, p. 43), the Lockport and Niagara Falls Railroad is shown leading into the end of what would become Bridge Street (the street leading to the Goat Island Bridge). The post office and two of General Whitney's hotels, the Cataract and the Eagle were located near the terminus of the line.

By 1856, these hotels are shown isolated and bypassed by the line of the New York Central, which made a continuous line of its run from Buffalo and points east, curving north to travel along the rim of the gorge. A new cluster of hotels arose around the New York Central Depot, the Franklin Hotel, the Empire and the Frontier Mart, own by the Porter family (map, reprinted by Adams, 1927, p. 48). The Porters had also extended the canal established in the 1830's down toward Prospect Park to parallel Canal Street, and established yet another, shorter canal, closer to the falls and cutting through Bridge Street. The great hydraulic canal extending from Port Day to the Canal Basin was also shown on the map.

Loker (1963) attributed the interest of Samuel DeVeaux in installing the Lockport and Niagara Falls Railroad to development of Niagara Falls as a tourist center. DeVeaux "sold some of his land for the railroad right-of-way. This road, The Lockport and Niagara Falls Railroad, referred to locally as the Strap Railroad, had Judge DeVeaux as one of its Directors" (Loker, 1963). DeVeaux promised financial backing if the bed of the tracks "followed the Gorge from the approximate area of the present Whirlpool Rapids Bridge to the station near The Falls on Falls Street" (Loker, 1963). The other Directors wanted land east of this, because it was cheaper, but DeVeaux threatened to withdraw all his support if they did so. DeVeaux succeeded. "... the first trains were advertised as "Two trips per day by steam power; twenty-four miles in about one hour and forty minutes" (Loker, 1963).

Associated with the railroad, whose bed appears to have followed the course of the gorge rim, but then angled away from the gorge forming the hypotenuse of a triangle of land west of the tracks, which became DeVeaux College and the tourist concession where tourists could look down at the Whirlpool. Thus isolated form the rest of the community, a fine old woods on this property was and has been up until the present, generally preserved form development.

DeVeaux may or may not have built a hotel in this general area, the Monteagle Hotel. In his will, signed the day of his death on August 3, 1852 (Loker, 1963, p. 25) he stipulated that "the Mount Eagle property may also be retained for the institution [DeVeaux College], or sold for its benefit ..." (Loker, 1963, p. 27). Loker indicated that "the hotel on this property, known as the Monteagle Hotel, built between 1848 and 1855, was torn down in July of 1936. The site today is a shopping plaza, starting at 2645 Main Street, Niagara Falls. The property itself, many acres reaching to the North and East of the Hotel, was sold over the years; the greatest single sale was in the 1890's for the development of the area known as McKoon Venue today for housing. The Hotel was discussed in "Passing of Famous Niagara Hostelry" by E. T. Williams in the Niagara Falls Gazette, Thursday, July 30, 1936" (Loker, 1963, pp. 168-9).

Loker indicated that "the first major sale did not take place until April of 11856 and this consisted of land to the North of the [DeVeaux School] lots," land for what would become Niagara University. Yet the hotel was built on the Mont Eagle property prior to 1856. On March 8, 1855, the first locomotive crossed Roebling's Niagara Railway Suspension Bridge, the permanent bridge that replaced Judge DeVeaux's 1848 suspension bridge designed by Charles Ellet, Jr. The Monteagle Hotel "opened near the bridge, and became a well-known landmark until it was demolished in 1936" (Greenhill, 1984). If we do not assume the trustees of DeVeaux College developed this hotel, then the property was sold to private interests within a year of two of DeVeaux' death.

In 1853, "the great New York Central Railroad was put together through a massive merger of tiny railroads across the state ... Buffalo was the New York Central's western terminus" (Brown & Watson, 1982), including, in 1853, the Lockport and Niagara Falls Railroad as well as the Rochester, Lockport and Niagara Falls Railroads (Scott & Scott, 1983). In 1855, the Niagara Railroad Suspension bridge was built making connections into Canada and used jointly by the New York Central and the Great Western Railways (Greenhill, 1984). The Central's Lewiston lines opened in 1854 (Scott & Scott, 1983, p. 34) and a perhaps modified map made for the US Lake and Topographic survey of an indeterminate year, reproduced by Spencer (1895), shows the Lewiston Branch of the New York Central. On this map, just to the east of this line were the rails of the Rome, Watertown and Ogdensburg Railroad.

The curious item about this map is that it appears as though the Lewiston line was built down at the base of the Niagara gorge, descending at a point north of Devil's Hole and the outlet of the Bloody Run creek into that cove in the east wall of the gorge. Naturally, if the line extended to Lewiston and the steamship and ferry docks there, opposite Queenston, Ontario, it had to descend the escarpment. The grade, for example, of the Erie and Ontario Rail Road in the 1830's coming up the escarpment from Queenston, Ontario, before heading south was from 3 to 5 degrees (Seibel, 1990, p. 100). This railroad chose to construct this grade in the east-west orientation of the Niagara Escarpment. It appears as though the New York Central built their grade in a north-south direction in the gorge itself at the north end where the trains could ascend or descend on the grade needed where suitable to the capacity of the steam engines. The remains of two railroad beds are still evident in the gorge near Artpark.

The Rome, Watertown and Ogdensburg Rail Road does not appear to have connected with Lewiston on the Lake Plain but to have carried its passengers east of the Lewiston line and along the rim of the gorge, turning east just before the escarpment and descending it on a long east-west grade, much like the Erie and Ontario Railroad across the river. By 1921, the Rome, Watertown and Ogdensburg was a division of the New York Central (map, Adams, 1927, Vol. 1, after p. 51), and the Lewiston line was still operational.

APPENDIX 3. OWNERSHIP OF GOAT ISLAND
AND ADJACENT RIVER MARGIN

1. Seneca Indian Nation — In 1763, Senecas attacked a British party at Devil's Hole along the rim of the Niagara Gorge south of Lewiston.

a. One year later, the Senecas ceded to the British by treaty between Great Britain and "nearly all the Indian tribes of North America" (Porter, 1900) held at Fort Niagara "all the land along the Niagara river, four miles wide averaging two miles in width on each side thereof, from Lake Ontario to Lake Erie' excepting the islands in the Niagara River. Due to a lack of cooperation by the Seneca at this meeting, and after British threats to send "Bradstreet's army" to ensure their participation, the Senecas made a conciliatory offer to cede to Sir William Johnson personally the river islands. Johnson in turn offered these to the British Crown.

b. A probably spurious claim by John Stedman that the same Indians "gave him all the land between the Niagara river" and Fort Schlosser, upriver east of the eastern boundary of the present Niagara Reservation, "some five thousand acres" (Porter, 1900). The State rejected the Stedman claim by 1823.

2. English Ownership and The Mile Strip — Details of British ownership and colonial rights are described below in the context of land claims subsequent to the American Revolution:

"Massachusetts traced its claim to the original [British] charter. On November 3, 1620, King James I had granted to the Plymouth Company title to all the land between the fortieth and the forty-eighth parallel that extended from sea to sea. A portion of this territory was ceded by the Plymouth Company to the Massachusetts colony in 1628 and was confirmed a year later by King Charles I. The charter of 1628 was vacated in 1684, and another one was granted by William and Mary in 1691. This charter included the lands from 42°2' to 44°15' that ranged from the Atlantic to the Pacific Ocean. Western New York extended south to the 42° and almost reached the 44° to the north. On this basis, Massachusetts claimed the land in western New York."

New York had a rival claim based on the decision of King Charles II, in 1664, to give James, Duke of York, and his brother, the Dutch possession of New York "including the present State of New Jersey. The land extended from the west side of the Connecticut River to the east side of the Delaware River. The royal grant was later confirmed by treaty with Holland. The Duke of York thus acquired land which was also claimed by Massachusetts."

New York, in 1780, and Massachusetts, in 1785, both ceded their claims to the United States. The 19,000 square miles east of the present boundary of western New York, however, was claimed by Massachusetts founded on "her claim under the royal charter of 1628, while New York claimed the land by virtue of her protectorate over the Iroquois Six Nations." In 1786, "New York acknowledged the right of the Commonwealth of Massachusetts to preemption of the soil, that is the right of first purchase from the Indians of all land in western New York except for a mile-wide tract along the Niagara River" (= The Mile Strip], a tract comprising 6 million acres. "In return, Massachusetts recognized the political sovereignty of the State of New York over this same area." The Mile Strip, originally sixteen miles in length stretching from Lake Ontario (Adams, 1927), was "reserved from cession [by New York State] to Massachusetts its title to this land" in 1786 (Adams, 1927). The Mile Strip was reserved to the State of New York specifically as a provision of the Treaty of Paris (Adams, 1927), and may have been established for military reasons.

3. Ownership by the United States — Subsequent to the American Revolution, which ended by the Treaty of Paris, 1783, with Great Britain and as a tribute to Indian rights, perhaps granted due to favorable Indian alliances during that conflict, English possessions in what is now western New York State were legally given to the aboriginal (pre-European) nations traditionally residing there, although the Indians, during the American Revolution, "suffered a crushing defeat in the Clinton-Sullivan campaign" (Chazanov. 1970).

During the Revolutionary war, the Iroquois strength was attributed to "their peculiar family and tribal organizations, and to close union and combined action, in all their warlike operations. General Sullivan's expedition had been sent out to chastise these hostile tribes for their barbarous conduct at Wyoming and Cherry Valley and elsewhere, in 1778, and to prevent future incursions ... overcoming all resistance, driving the Indians from their villages and utterly destroying all their dwellings and crops in the Genesee Valley, on both sides of Seneca and Cayuga lakes, and in all that region. The Indians continued the struggle for two or three years longer, but with the close of the Revolutionary War all organized and open hostility ceased" (Porter, 1875). With their tribal infrastructure disintegrated, it was only a formality to clear away the Indian populations of western New York, with their legal rights traded for cash, for the sale and settlement of their lands.

After the American Revolution, Robert Morris bought the land west of the Genesee River, and sold it to Theophile Cazenove between 1792 and 1793 as agent for six Dutch banking houses who, in 1795, merged their interests to form the Holland Land Company. This arrangement was made indifferent to the fact that the Indian nations still had legal title to these lands. The Indian title had to be terminated by Morris, and this was effectively done, except for the Tuscarora, Tonawanda, Buffalo Creek, Cattaraugus and Allegany Reservations, at the Treaty of Big Tree in 1797. Apparently the islands in the Niagara River were also exempted from American ownership after the 1797 treaty, as was the Mile Strip along the Niagara River.

4. Ownership by New York State — After settling its sovereignty over land in its western territory, New York surveyed and sold lots along the Mile-Strip, which had been "reserved to the State out of the Cession to Massachusetts in 1786" (title of a map of lots sold reproduced by Adams (1927) and in part by Scott and Scott, 1983. The map was originally drawn by Joseph Annin in 1805 (Adams, 1927). Indian rights, however, still prevailed on this land, and a treaty with the Seneca was made in 1802 (Adams, 1927), but still apparently exempting the islands in the Niagara River from State ownership. Eventually the "Mile Strip" became 36 miles in length (Adams, 1927). The Indian title to all the islands in the Niagara River had finally been extinguished by the State of New York "only a few weeks before" October, 1815 for "$1,000 cash and $1,500 a year in perpetuity" (Porter, 1900).

The Mile Strip was surveyed by the State in 1798 and lots were sold at auction on February 26, 1805. Purchasers and the lots they bought were given in the Annin map of 1805 (note that, probably as an oversight, Goat Island is not drawn on this map although the other major river-islands upriver were). Purchases by the brothers Porter, Benjamin Barton, Joseph Annin and associates reflect the anticipated loci of the Porter's dockage and portage sites, and riparian rights to land just above the American Falls. According to the Annin map, land on the mainland now in the Niagara Reservation State Park included lots 43 and 44 (the old Stedman farm), purchased by Joseph Annin, uncle to Benjamin Barton, business partner to the Porters. The deed for Goat and other islands in the Niagara River was granted to Augustus Porter in 1816. The State repurchased by condemnation these properties by 1885.

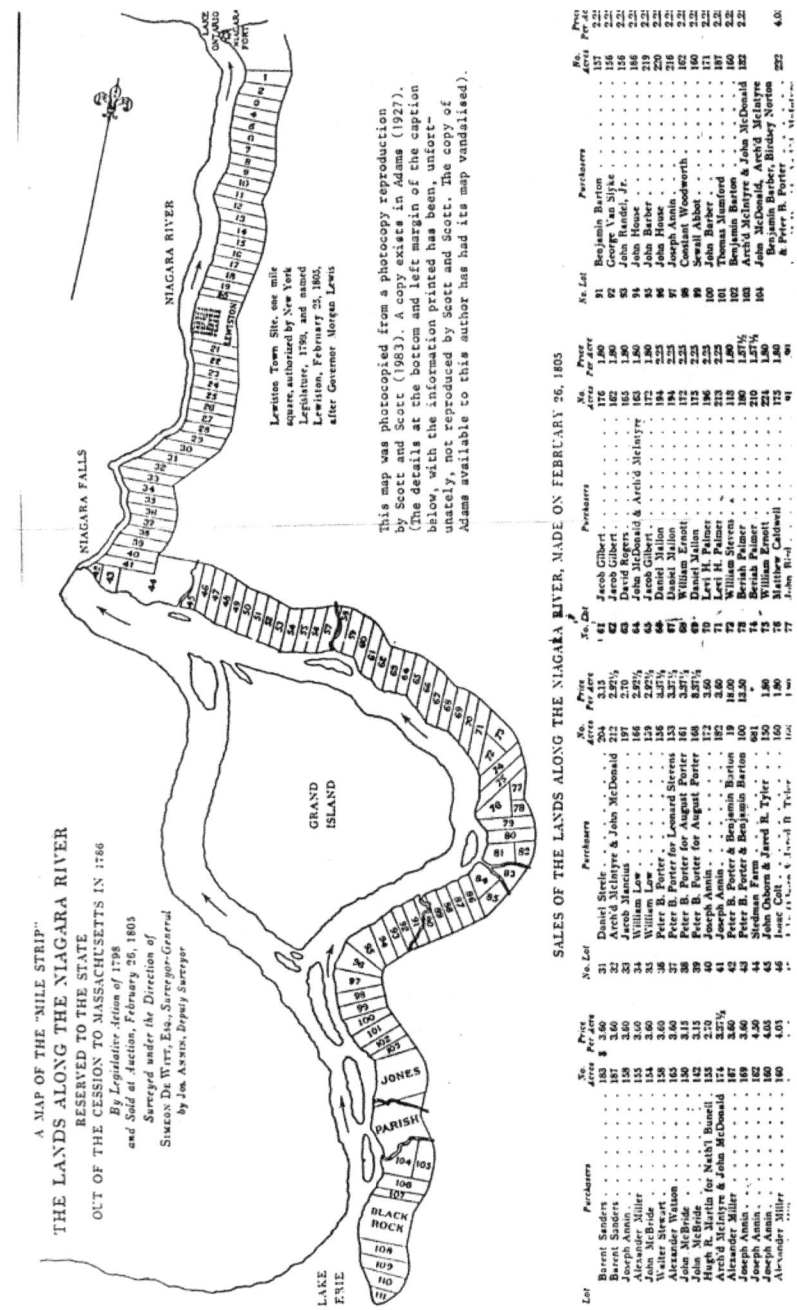

A MAP OF THE "MILE STRIP"
THE LANDS ALONG THE NIAGARA RIVER
RESERVED TO THE STATE
OUT OF THE CESSION TO MASSACHUSETTS IN 1786
By Legislative Action of 1798
and Sold at Auction, February 26, 1805
Surveyed under the Direction of
SIMEON De WITT, Esq., Surveyor-General
by Jos. ANNIN, Deputy Surveyor

Lewiston Town Site, one mile
square, authorized by New York
Legislature, 1798, and named
Lewiston, February 25, 1805,
after Governor Morgan Lewis

This map was photocopied from a photocopy reproduction
by Scott and Scott (1983). A copy exists in Adams (1927).
(The details at the bottom and left margin of the caption
below, with the information printed have been, unfortunately, not reproduced by Scott and Scott. The copy of
Adams available to this author has had its map vandalised).

SALES OF THE LANDS ALONG THE NIAGARA RIVER, MADE ON FEBRUARY 26, 1805

Appendices

The Mile Strip — This is the top portion of the Mile Strip Map in Adams (1927).

APPENDIX 4. LAKE LEVELS

My essay: "Some Potential Impacts of Conditions in the Saint Clair River on NYPA Relicensing, Lake Erie, and the Niagara River" has been recently posted on line:

http://www.mobot.org/plantscience/resbot/Niag/LakeLevels/StClairRiver.htm

In this essay, I mention glacial rebound, isostatic rebound in the Great Lakes region, as a possible explanation for some of the effects of dredging in the Saint Clair River on upper lake levels. This idea was intriguing enough for me to consider additional thoughts on the rebound issue and how it might play itself out elsewhere in the Great Lakes system, such as downstream of Lake St. Claire.

There are at least two paradigms to bear in mind when viewing water level issues, one is that atmospheric weather patterns are changing, average annual temperatures are rising and that generally things are becoming more arid. Lowered lake levels, according to this scenario, may be attributed to reduced moisture entering the watershed.

Another scenario to account for lowered lake levels is the isostatic rebound paradigm, here used to explain to some extent the increase in erosion in the Saint Clair River. If I am not mistaken, the model is that in the Laurentian region of Ontario/Labrador where the nucleus of the great ice sheet of the last "ice age" originated, around Hudson Bay, the land affected by the ice sheet development, was depressed under the weight of this mass of ice. Softer, hotter, molten rock upon which the upper geologic strata "float", yielded to this weight. With the final melting of this ice sheet, slowly the brittle, cold upper layers are rising in elevation relative to sea level.

The effect on the hydrology of the Great Lakes Watershed is to increase the stream gradient on north to south trending flow patterns through streambeds, such as the Saint Clair and the Detroit rivers. Bodies of water would have their lakebeds tilted, the north shores rising relative to the south shores in these basins: Lake Superior would tip its water mass toward the southern shore, presumably putting hydrostatic pressure on the strait [see note at end of essay] of Saint Mary, or the Saint Mary's River with possible augmented stream flow through that river through the Sault Saint Marie.

In Lakes Michigan and Huron, both basins elongated in a north-south direction, the water mass would press upon the southern parts of the basin: shorelines would tend to become drowned at their southern ends, more exposed at their northern, hence there should be a decrease in hydrostatic pressure through the Straits of Mackinac. This hydrostatic pressure would stimulate erosion regimes through increased gradient as well, in the straits between Lake Huron and Lake Erie.

The south shore of Lake Erie and Ontario would become drowned relative to the more exposed north shore of both east-west trending lake basins.

The strait between Lake Erie and Lake Ontario is the Niagara River. The curious thing is that this strait, also oriented on a north-south axis like the Saint Clair and Detroit rivers upstream, has its flow character reversed relative to those of the Lake Huron and Erie corridor: the current flows north.

If isostatic rebound accounts for increased hydrostatic pressure by shifting of the water mass in Lake Huron (whose levels may also be determined by increased flow through the Saint Mary's strait from the rise of the north shore of Lake Superior) and from increased stream gradient, then the reverse must be true in the Niagara River.

Stream gradient must be reduced in the Niagara River because its northern end at Fort Niagara, N.Y. and Niagara-on-the-Lake, Ont., has risen relative to its foot at Buffalo, N.Y. and Fort Erie, Ont. Volume in the Niagara River might also be decreased according to the isostatic scenario, because hydrostatic pressure by the volume of water in Lake Erie has declined, due to the retreat of this mass to the southern shores of the Lake, where all the U.S. towns are located, like Cleveland, Ohio.

Furthermore, due to the existence of Niagara's handsome cataracts, the backwater that should be swelling the volume of the water in the River north of the cataracts (called the Lower Niagara River) due to augmentation of the southern shore of Lake Ontario is essentially dammed up at the base of the Falls in the plunge pool area—an accumulation of water useless for scenery and for hydroelectric power generation.

Unfortunately, the cataracts themselves would be rising relative to their elevation at Buffalo-Fort Erie. That this process has a long history can be seen in the number of abandoned gorges or ravines that litter the north-facing Niagara Escarpment through New York State and Peninsular Ontario. These old "spillways" are the remnants of north-flowing surface streams that no longer function as streams, their headwaters probably now the source of wetlands south of their outlets, or the victims of stream capture.

The curious development of islands at the mouths of tributary streams flowing east-west into the north-south Niagara River axis may be due to isostatic rebound contributing to a decrease in gradient at the mouths of these streams, forcing the streams to be deflected at their mouths. Isostatic rebound could also be a factor in the characteristics of stream depth relative to the northern and southern shorelines and island boundaries in the Niagara River as they trend in an east-west departure from the north-south axis of most of the river, in the area between the head of Grand Island and the brink of the cataracts.

At Niagara, the water is so shallow at the brinks of the cataracts, especially in the central and north-central Grass Island pool area where the two channels (Chippawa and Tonawanda) rejoin after flowing around Grand Island, N.Y., that any increase in elevation in the northern area of the stream bed must contribute to a deterioration in the hydrostatic force necessary to ensure the kind of water levels most useful to the two hydroelectric power facilities that divert water from the head of Grand Island. The demands of flow for scenic purposes on which the casinos and their government partners rely for their tourist customers will eventually conflict with diversion demands just upstream from the cataracts.

Isostatic rebound happens throughout the Great Lakes Watershed region, contributing to enhanced elevations in two additional watersheds, that of the Mississippi River in the United States where improvements in stream gradient are probably beneficial, but also in the Red River watershed in Canada that channels water into Hudson's Bay where stream gradient would deteriorate.

Stream gradient would also deteriorate in the strait of Saint Lawrence, connecting Lake Ontario with the Atlantic Ocean, whose trend is from the southwest to the northeast.

Water should then be entrained in both Lakes Erie and Ontario, but since the gradient of the stream outlets of both lakes is being reduced by regional isostatic rebound. The water levels of these two lakes actually may be rising because of modifications in elevation of their present outlets.

The far future scenario using this paradigm is that eventually these watersheds will resume their use of the Mississippi spillway system and the area of the watershed for that river will be increased.

Lake levels and electric power production — In conclusion, use of the isostatic rebound paradigm to explain matters for increased erosion and flow through the Saint Claire and Detroit rivers perhaps must also explain the great urgency to get more water into the Lake Erie and Lake Ontario lake basins because, as erosive forces build in the southern outlet of Lake Huron, the forces of sedimentation are increasing in the

Appendices

Niagara River, the outlet of Lake Erie. Dredging the Saint Clair would have the benefit of perhaps maintaining adequate water volume in the Niagara and Saint Lawrence rivers upstream to generate hydroelectric power (in addition to navigation issues). This scenario may reinforce the suggestion of a link between government activities in both sections of the Great Lakes.

With the price of oil varying greatly at the present time, the northeastern United States and Canada is a vast though presently fickle market. The value of hydroelectric power, however, generated both by the Province of Ontario and the State of New York, can increase greatly.

Currently, the benefits of financial "resettlement" during the relicensing process of the New York Power Authority are expected to benefit the American side of the Niagara River in U.S. communities most directly associated with the Niagara River.

However, a case might be said that the distribution of resettlement money, derived from users of Niagara's hydroelectric bounty, should perhaps be distributed to a broader region of effect to include upstream states and communities in both the United States and Canada, especially if an energy crisis is imminent with resumption of the cold or heating season in years to come.

The Great Lakes together constitute a great river with a continuous shore- or coast-line with huge embayments (lakes) connected by narrow channels (straits). This huge river system flows eastward to the Atlantic Ocean, its final journey down the strait of the Saint Lawrence, ending in the gulf of the same name. The mixture of fresh water and salt water in this gulf should have a very interesting biology compared with either the Lakes or the Ocean themselves. A strait is a relatively narrow waterway connecting two larger bodies of water and the Great Lakes has quite a few, all locally called "rivers." A river, in contrast is "a stream of water bearing the waste of the land from higher to lower ground, and as a rule to the sea." (Dictionary of Geological Terms, according to the American Geological Institute). One can imagine the chemical and biological distinctions in the waters of the two systems, a strait and a river.

One of the permutations of the Niagara River (strait) is its division into two channels, the Chippewa Channel which occupies the western branch of the river as it divides and flows around Grand Island (N.Y,) and is wholly in the Dominion of Canada, and the Tonawanda Channel, the east branch that flows around the other side of Grand Island, and is wholly in the territorial United States. What is seldom understood is that the majority of the water volume in the Niagara River is contained in the Canadian part of the River and the sovereign boundary is not in the middle of the Chippewa channel, as one might expect, but almost up onto the beach on Grand Island (NY), maximizing Canada's claim to the hydroelectric potential at the cataracts just downstream. The reason the Canadian Falls is called Canadian, and the American the American Falls is due to the fact that one cataract resides in Canadian territory, the other in the United States. The Canadian cataract (usually referred to as the Horseshoe Falls) has most of the water volume, that of the American Falls is so poorly hydrated that "control structures" must be built just upstream of Goat Island to divert water over the American Falls for visual reasons. If one scrutinizes topographic maps for Goat Island (USA), one can see the Canadian territorial boundary seems to actually cut across Goat Island at Terrapin Point, which famous promontory may actually reside in the Dominion of Canada.

Boundary issues — These hydraulic follies arose out of the boundary settlements of the Treaty of Ghent signed in Europe after the ambiguous ending of the War of 1812 where the engineers of the British Army, who had designs on and had studied the hydraulic potential of the cataracts all during the colonial era, perhaps had the last laugh as they withdraw behind the new frontier, taking the water with them. The U.S. Army Corps of Engineers still must endlessly dredge the Tonawanda Channel, whereas the Chippewa Channel seems to still exist in its aboriginal character.

A small lake could be said to be formed where the confluence of the Chippewa and Tonawanda Channels of the Niagara River meet just above the cataracts, where the river is again divided by an island (Goat Island, N.Y.). This small lake is called the Grass Island Pool. Grass Island, alas, like so many charming little islets that occurred on both shores above the Cataracts, such as Cedar Island and Hogg Island, was joined to the mainland as a result of water diversion that resulted in these islands having no water in their channels anymore and, hence, no claim to island status. Presently, as more and more water is diverted, there are now more islands above the brinks in the stream channels than there have ever been in the history of the cataracts—so many that no one bothers to give them names.

The water level in the Grass Island Pool determines, by international treaty, how much water is diverted to the Canadian or the American hydroelectric plants downstream, and how much goes over the brinks to amuse tourists. The diversion structures of both countries are located on the shores of the Grass Island Pool.

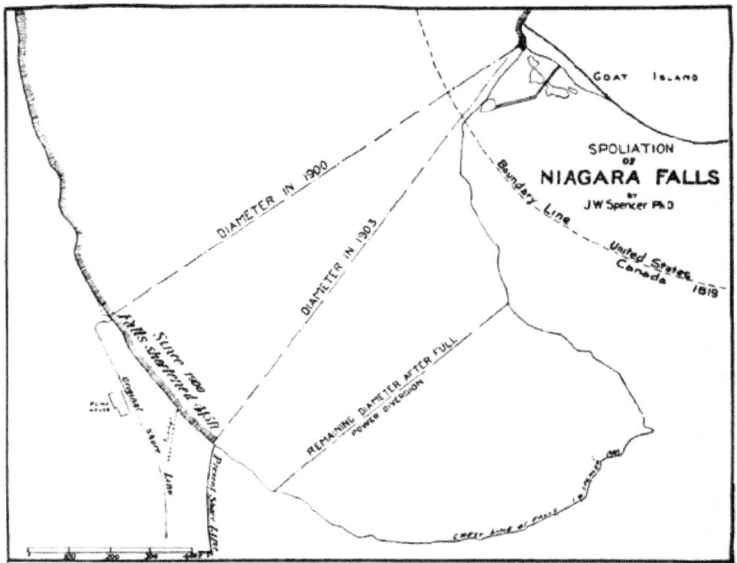

Spoliation map — Spencer (1908) shows here the 415 feet of Horseshoe Falls eliminated on the Canadian (left) side, while much later Goat Island was extended to the Boundary line by fill.

Spencer — The problem of water diversion, hydroelectric power production and the integrity of the natural scenery at Niagara Falls has been discussed and fought over since early times. In 1907, J. W. Spencer (1850–1921) published a geological survey of Niagara Falls (Spencer, 1907), extended in 1908 (Popular Science Monthly) into a concern for water available as a scenic attraction at Niagara Falls given the great demand for water power upstream. He wrote:

"International Waterway Commission established for saving the Falls. — Before this time, the late Honorable Andrew H. Green had secured the passage through Congress of a bill, authorizing the establishment of an International Waterway Commission, his specific object being the preservation of Niagara Falls. Indeed it was for this same object that the International Park at Niagara had been established at an earlier date, largely through the efforts of the Earl of Dufferin, Governor General of Canada, and Mr. A. H.

Green, of New York, who subsequently prevented the intrusion of all power structures in the state reservation on the New York side, a policy unfortunately not followed by the government on the Canadian side. Here even the park was widened, at the cost of the falls, in curtailing their crest-line by several hundred feet. Yet among those interested in the power companies it was commonly said that they were improving the park; a few, who were powerless, seeing through this sophistry.

"I found that the outlets of both Lake Erie and Lake Ontario had been recently [through natural forces] lowered, while Mr. Thomas Russel, of the U.S. Lake Survey, had previously made the great discovery that the outlet of Lake Huron had also been lowered. This was the starting- point of the investigation into the spoliation of Niagara. The channel of the river had been deepened just after 1890, owing to natural scour by the currents, the effectiveness of which was increased by the powerful jamming of vast quantities of ice against the barriers at the Upper Rapids, immediately above the falls, and to a small extent by the shifting of the boulders on the river bed just below the outlet of Lake Erie."

Spencer found and well documented the fact that the levels of the Great Lakes varied greatly, and the franchises of the great power companies would deplete the water going over the Falls during periods of low water. He pointed out that the Falls "begin as the water passes over a rim of rock ... which crosses the river at the head of Goat Island. This is the 'critical point,' not merely in the distribution of water over the falls, but also in the level of Lake Erie, and indirectly of Lakes Huron and Michigan." He pointed out that more than 400 feet of the Horseshoe Falls had already by cut off on the Ontario side apparently to eliminate shallow water, while further diversion of water not accounting for variation in Lake Erie water levels would dewater during low water the area of Goat Island next to the Horseshoe Falls. In fact, today about 300 feet of the Falls has been cut off by deposition of fill such that the boundary of Canada now touches Goat Island. This elimination of about 700 feet of the crestline of Horseshoe Falls, a shortening of 2950 feet in 1900 to 2250 feet today, is doubtless due to diversion of water upstream in the intakes for hydroelectric power.

Spencer (1908) indicated that "full diversion" to the limits of the franchises of the power companies would reduce the width of the Horseshoe Falls greatly, but legally the power companies have been prevented from full diversion. He warned, however, that:

"Any attempt at restoring either the American or the eastern side of the Canadian Falls, by deepening the channels on that side of the river, would increase the velocity of the currents above and cause an extraordinary demand on Lake Erie, the result of which would be the lowering of its level at an enormous cost. The same physical changes would subsequently take place in the Huron outlet as a consequence of the lowering of the Erie level. The artificial deepening of the channel would also increase the scour, not merely of the Niagara River, but also the St. Clair channel, which lies in deposits of sand and clay."

One should note that Spencer was writing in 1908. His opinions are based on thorough-going research. He was a pioneer Canadian geomorphologist, earning his doctorate at Göttingen, Germany. He taught at King's College, Windsor, Ontario, and in 1882, became professor and Director of the Museum of Natural History at the University of Missouri (Middleton, 2004).

In the 1900's the St. Clair River was dredged to 22 feet, and again in the 1930's to 25 feet (Egan, 2013). In 1964, the Army Corps of Engineers dredged the outlet of the river into Lake Erie, this time to 27 feet. This was about the time they also dredged the area of Niagara River at the power inlet on the U.S. side. The St. Clair River channel subsequently was so deeply scoured that the water level of Lake Huron was affected. This work was only uncovered when residents of the north shore of Georgian Bay hired a private investigation firm to find out why the boats and piers of their summer houses were stranded on dry land by the now low water level. It is hard to imagine that the Corps was ignorant of the soft consistency of the river bottom and its tendency to scouring, as described above by Spencer. The levels of the Great Lakes since 2000 have dropped consistently, possibly associated with increased evaporation caused by lowered albedo due to less ice cover (Egan, 2013). This makes Spencer's cautions now more relevant than ever. The control of the level of Lake Erie is critical to maintaining electric power production in New York State and Ontario. Perhaps for such a reason, this most shallow of the Great Lakes has a fairly steady water level.

APPENDIX 5. THE NIAGARA FALLS MUSEUM

Thomas Barnett's Niagara Falls Museum, built in 1827 on the Canadian side just below the Horseshoe Falls, was an example of a well-run and interesting tourist attraction. As a spectacle for tourists, at one point in a commercial battle with a competitor, he staged an Indian Burial and, in 1872, a Buffalo Hunt, an extravaganza "employing a large number of Indians, who were brought from the American Mid West under the guidance of Wild Bill Hickock and Buffalo Bill Cody" (Kiwanis Club, 1984).

To appreciate some idea of the rowdy free-for-all of the cities at Niagara Falls of the time, an extensive description of the Buffalo Hunt is here quoted from O'Connor (1959):

"It took two weeks for Hickok and his assistants, hauling the buffaloes along by ropes, to get them to the railroad yards at Ogallala and load them into cattle cars. Everybody who witnessed the spectacle of Hickok cramming buffalo into railroad cars naturally thought he was a madman.

"Hickok next recruited four hungry but fairly tame Comanches as members of the company. One of the Comanches had a pet cinnamon bear, another somehow had acquired a monkey, neither of which they intended to leave behind. Hickok allowed the Indians to bring along their pets.

"Hickok & Co. set out for Niagara Falls late in June—the impresario himself, three cowboys, four Comanches, six buffaloes, a monkey, and a bear.

"The show opened July 20, with four or five thousand people gathered around the wire enclosure in which "The Daring Buffalo Chase" was to be performed ... .

"It was, on the whole, an exciting performance, although it did not proceed as planned. In the first place the buffaloes were reluctant to be chased; they had to be pushed into the arena and then just stood there, balking ..., snorting and pawing the ground occasionally. Finally Hickok fired a gun and they took off in a wild charge. Round and round the enclosure they went, with the Comanches in pursuit. A few minutes later a pack of dogs attracted to the scene by the excitement, an equally joyous mob of small boys, and a number of adult spectators entered the enclosure and joined the chase...."

"Then the situation got completely out of hand. With a concerted charge the buffaloes tore through the fence and thundered through the nearby residential section. Somebody unbarred the bear's cage to the delight of both the bear and the crowd .... The monkey took this opportunity to break out of his cage, climb to the top of a wagon, and hurl everything he could lay his paws on at the crowd."

"To cap it all, when the dust had settled on this exhibition an English tourist came up to Hickok and inquired, 'I say, my good man, are you an Indian or a white man?'

"Hickok knocked him flat, saying, 'That's the kind of man I am,' and fled to the nearest saloon.

"That wasn't quite the end of Hickok's misfortunes. The Comanches were clamoring to be sent home, although Hickok was flat broke. He finally managed to sell the six buffalos to various Niagara Falls butchers. The proceeds were used to buy tickets home for the Comanches. Hickok and the cow hands were forced to beat their way back to Kansas City on the freights."

ACKNOWLEDGEMENTS

Donald E. Loker, librarian at the Niagara Falls Public Library provided assistance with the history of the Niagara Reservation. Richard H. Zander, Curator of Botany at the Buffalo Museum of Science and later Research Associate at the Missouri Botanical Garden, provided valued support, much-appreciated encouragement, and critical editorial and computer expertise. Various officials associated with the New York State Department of Parks and Historic Preservation gave various necessary access and permissions over the years; I particularly thank Mario Pirastru, then Niagara Regional Director, Thomas B. Lyons, Director of Resource Management, and State Parks Commissioner Rose Harvey. Cynthia Van Ness, Director of the Library and Archives at the Buffalo and Erie County Historical Society was ever helpful in search of important documents in the extensive holdings of that institution.

BIBLIOGRAPHY

Adams, Edward Dean. 1927. Niagara Power: History of the Niagara Falls Power Company 1886-1918. 2 Vols. Niagara Falls Power Company, Niagara Falls, New York.

Adamson, J. E. 1985. Niagara, Two Centuries of Changing Attitudes, 1697–1901. Corcoran Gallery, Washington, D.C.

Albright, H. M. (as told to R. Cahn). 1985. The Birth of the National Park Service. Howe Brothers, Chicago.

Beveridge, C. E. "Planning the Niagara Reservation" pp. 16–25 in The Distinctive Charms of Niagara Scenery: Frederick Law Olmsted and the Niagara Reservation. 1985. Buscaglia-Castellani Art Gallery of Niagara University, Niagara Falls, New York.

Bridgwater, W. & S. Kurtz. 1963. The Columbia Encyclopedia, ed. 3. Columbia University Press, New York.

Brown, R. C. & B. Watson. 1981. Buffalo, Lake City in Niagara Land. Windsor Publications, Inc., Buffalo.

Buck, R. S. 1898. The Niagara Railway Arch. Amer. Soc. Civil Engineers 40: 125–150.

Chazanof, William. 1970. Joseph Ellicott and the Holland Land Company; the opening of western New York. Syracuse University Press, Syracuse, New York.

Dow, C. M. 1921. Anthology and Bibliography of Niagara Falls. Vols. 1 & 2. State of New York, J. B. Lyon printers. Albany.

Eckel, P. M. 2013. Botanical Heritage of Islands at the Brink of Niagara Falls. Botanical Services, St. Louis: CreateSpace Independent Publishing Platform, Amazon.

Egan, D. 2013. Once-steady Great Lakes' flow altered by dredging, dams and now warming temperatures. Also: Does Lake Michigan's record low mark beginning of new era for Great Lakes? Also: Water flushes through a greatly widened drain below Great Lakes Michigan, Huron. Milwaukee Wisconsin Journal Sentinel July 27, 2013 http://www.jsonline.com/news/wisconsin/once-steady-great-lakes-flow-altered-by-dredging-dams-and-now-warming-temperatures-217150821.html

Fox, A. M. 1986. Designated Landmarks of the Niagara Frontier. Meyer Enterprises, Buffalo.

Gardner, J. T., Director. 1880. New York State Survey. Special report on the preservation of the scenery of Niagara Falls, and fourth annual report on the triangulation of the state for the year 1879. Albany: Charles Van Benthuysen and Sons, pp. 27–31.

Goldie, John. 1819. Diary of a Journey Through Upper Canada and Some of the New England States, 1819. Privately published. In the copy seen from the Sidney B. Coulter Library, Onondaga Community College, Syracuse, New York, 13215, penciled in the title page is the notation "Toronto, Ontario, 1961." Mr. Willman Spawn of Philadelphia, Pennsylvania, has written an introduction and notes to the latest published edition.

Goldman, Mark. 1983. High Hopes: The Rise and Decline of Buffalo, New York. University of the State of New York Press, Albany.

Grabeau, A. W. 1901. Guide to the Geology and Paleontology of Niagara Falls and Vicinity. Bulletin of the Buffalo Society of Natural Sciences. Vol. VII, No. 1. University of the State of New York, Albany.

Green, E. [no date]. The Niagara Portage Road. (Reprinted from the Ontario Historical Society's Papers and Records, Vol. XXIII) cited by Way, 1946.

Green, S. S. 1905. Andrew Haswell Green, a Sketch of his Ancestry Life and Works. Proceedings of the American Antiquarian Society. New Series Vol. XVI. Oct. 1903—Oct. 1904. Worcester, Massachusetts.

Greenhill, R. 1984. Spanning Niagara: The International Bridges, 1848–1962. Univ. Washington Press, Seattle.

Gurney, J. J. 1841. A Journey in North America, Described in familiar letters to Amelia Opie. Private circulation, Norwich, pp. 322–323 [cited in Dow, 1921].

Hall, J. 1882. Geology of New York. Part IV comprising the survey of the fourth geological district. Caroll & Cook, Albany, New York. Partially reprinted in the Eighth Annual Report of the Commissioners of the State Reservation at Niagara for 1890–1891, 1892. Albany.

Hamilton, G. H. 1943. Plants of the Niagara Parks System of Ontario. Ryerson Press, Toronto.

Hennepin, F. L. 1697. A New Discovery of a Vast Country in America, extending above four thousand miles, between New France and New Mexico: with a description of the Great Lakes, Cataracts, Rivers, Plants and Animals: also the Manners, Customs and Languages of the several Native Indians: and the advantages of commerce with these different nations, &c." Utrecht. [London, 1698].

Ingraham, J. W. 1834. A Manual for the use of Visitors to the Falls of Niagara, intended as an epitome of, and temporary substitute for, a larger and more extended work, relative to the most stupendous wonder of the world. Charles Faxon, Buffalo, New York.

Kowsky, F. R. 1985. The Distinctive Charms of the Niagara Scenery: Frederick Law Olmsted and the Niagara Reservation. Catalogue for 1985 Niagara Reservation art exhibit.

Loker, D. E. 1963. A History of DeVeaux School, 1853–1953. Fose Printing, Niagara Falls, New York.

Bibliography

Lyell, C. 1845. Travels in North America, in the years 1841–2, with geological observation on the United States, Canada, and Nova Scotia. 2 Vol. Wiley and Putnam, New York.

Middleton, G. V. 2004. J. W. Spencer (1851–1921): his life in Canada, and his work on preglacial river valleys. Geoscience Canada 31. Online version: URL: https://journals.lib.unb.ca/index.php/GC/article/view/2750

Mizer, H. B. 1981. Niagara Falls. A Topical History, 1892–1932. Occasional Contributions of the Niagara County Historical Society, No. 24.

Morden, J. C. 1938. Falls View Bridges and Niagara Ice Bridges. F. H. Leslie, Ltd., Niagara Falls, Ontario.

Newton, N. T. 1971. Design on the Land. The Development of Landscape Architecture. Harvard University Press, Cambridge. The Niagara Book, a Complete Souvenir of Niagara Falls. 1893. Underhill and Nichols, Buffalo.

O'Connor, R. 1959. Wild Bill Hickock. Doubleday, New York.

Olmsted, F. L. 1880. In Gardner, J. T., Director. 1880. New York State Survey. Special report on the preservation of the scenery of Niagara Falls, and fourth annual report on the triangulation of the state for the year 1879. Albany: Charles Van Benthuysen and Sons, pp. 27-31.

Olmsted, F. L. & C. Vaux. 1887. General Plan for the Improvement of the Niagara Reservation, in Supplemental Report of the Commissioners of the State Reservation at Niagara. Third Annual Report of the Commissioners of the State Reservation at Niagara, for the year 1886. Pp. 9–50. The Argus Co., Albany, New York.

Otis, M. P. 1982. Draft Environmental Statement. Niagara Reservation Conceptual Master Plan. New York State Office of Parks, Recreation and Historic Preservation. July.

Parsons, H. A. 1836. The Book of Niagara Falls. Third Edition, with maps.

Petrides, G.A. 1958. A Field Guide to Trees and Shrubs. Peterson Field Guide Series. Houghton Mifflin Co. Boston.

Porter, A. H. 1875. Niagara from 1805 to 1875, by an old resident. Privately printed pamphlet in Buffalo and Erie County Public Library.

Porter, P. A. 1894. Historic Niagara, in the Tenth Annual Report of the Commissioners for the State Reservation at Niagara for the Year 1892-3. 1894. pp.8–71. Reprinted from the Niagara Book.

Porter, P. A. 1900. Goat Island, in Sixteenth Annual Report of the Commission for the State Reservation at Niagara for the Year 1899. Albany, pp. 75–129.

Promontory Partnership, The, EDAE Inc. Parsons, Brinckerhoff, Quade and Douglas, Inc. 1981. Niagara Reservation: Options for the Future, prepared for the New York State Office of Parks, Recreation and Historic Preservation.

Queen Victoria Niagara Falls Park. Appendix. Issue 10:pp. 1–60.

Recknagel, A. B. 1923. The forests of New York State. With an introduction by Liberty Hyde Bailey. Macmillan, New York.

Robinson, W. 1875. Alpine Flowers. John Murray, London, in Gardner, J. T., Director. 1880. New York State Survey. Special report on the preservation of the scenery of Niagara Falls, and fourth annual report on the triangulation of the state for the year 1879. Albany: Charles Van Benthuysen and Sons, pp. 27-31.

Roper, L. W. 1973. A biography of Fredrick Law Olmsted. Johns Hopkins University, Baltimore.

Rubbinaccio, M. 2013. New York's Father is Murdered! The Life and Death of Andrew Haswell Green. Pescara Publishing, Seattle, Washington.

Scott, S. D. & P. K. Scott. 1983. The Niagara Reservation Archaeological and Historical Resource Survey, 1983. New York Office of Parks, Recreation, and Historic Preservation. Historic Sites Bureau, March.

Seibel, G. A. 1985. Ontario's Niagara parks. 100 Years. A History. The Niagara parks Commission, Niagara Falls, Ontario.

Spencer, J. W. 1907. Falls of the Niagara: Their evolution and varying relations to the Great Lakes; characteristics of the power and effects of its diversion: Geological Survey of Canada, Publication 970, 490 p.

Spencer, J. W. 1908. Spoliation of the Falls of Niagara. Popular Science Monthly 73(October): 289–305. HathiTrust http://hdl.handle.net/2027/loc.ark:/13960/t7jq1b76g

Tiplin, A. H. 1988. Our Romantic Niagara. A Geological History of the River and the Falls. Niagara Falls Heritage Foundation, Niagara Falls, Ontario.

Todd, J. E. 1982. Frederick Law Olmsted. Twayne Publishers, Boston.

Van Cleve, A. H. 1903. Utilization of water power at Niagara Falls. Bull. Buffalo Soc. Nat. Sci. 8: 3–20.

Way, R. L. 1946. Ontario's Niagara Parks, A History. The Niagara Parks Commission, Niagara Falls, Ontario.

Welch, Thomas V. 1903. How Niagara was Made Free. The Passage of the Niagara Reservation Act in 1885. Publication of the Buffalo Historical Society 5: 325–359.

Wied-Neuwied, M. A. P., Prinz von. 1843. Travels in the interior of North America, with numerous engravings on wood and a large map; translated by H. Evans Lloyd. Ackman & Co., London, pp. 493–496.

Bibliography

Zander, R. H. 1976. Floristics and Environmental Planning in Western New York and Adjacent Ontario. Occasional Papers of the Buffalo Society of Natural Sciences 1: 1–47 (available online).

Zenkert, C. A. 1934. The Flora of the Niagara Frontier Region. Bull. Buffalo Soc. Nat. Sci. 16.

INDEX

Index

Cataract Bank, 152
Cataract Construction Company, 174
Cataract Hotel, 245
Cataract House, 30, 37
Cave of the Winds, 146
Cedar Island, 212
Central Park, 44, 88, 135
Chain Reserve, 40
Chesborough, 179
Chicago drainage canal, 195
chimney, 190
Chippawa, 35
Chippewa Channel, 267
Church, 16, 32, 49, 201
City of Niagara Falls, 227
Civil War, 41
Cleveland, 50, 82
Cliff Paper Company, 79
Clifton House, 35, 247
Clinton, 25, 59, 76
Cockburn, 3, 13
Commission of Appraisement, 83
Commissioners of the New York State Survey, 54
Commissioners of the Niagara Reservation, 90, 123, 210
Commissioners of the Queen Victoria Niagara Falls Park, 214
Commissioners of the State Reservation at Niagara, 214
conflict, 134
conifer, 3, 18
Cromley, 219
Crysler, 246
Daily Cataract, 172, 186
dangerous precedent, 198
Davies, 3
Day, 52, 144
Day Company, 31
degradation of the shoreline, 57
Delaware Park, 93
Department of the Interior, 226

Department of Tourism, 227
Department of Transportation, 227
DeVeaux, 233
DeVeaux College, 256
developers, 203
development, 163
Devil's Hole, 229
dining areas, 108
diversion limits, 157
diversion of water, 151, 163, 182, 270
diversity of trees, 118
Dorsheimer, 52, 143
Dow, 217
Dufferin, 53, 179
dynamiting, 115, 119
Eagle Tavern, 37
east meadow area, 125
Eastern Hemlock, 4, 13
ecological plan, 94
ecological restoration, 94, 112
ecosystem, 113
Edgerton, 124, 205
Edison, 68
educational standards, 230
Edwards, 18
electric lighting, 194
electric railway, 184
Ellicott, 26
Ely, 175, 193, 219, 237, 239
Emerson, 76
environment, 98
Erie & Ontario Railroad, 241
Erie Canal, 25, 240
escarpment, 59
Evershed, 178, 219
excavation, 167
faking nature, 115, 206
Falls View Station, 192
Farny, 17
federal government, 164, 199
fence, 112
Flint, 35

~ 299 ~

Lockport Water Supply Company, 156
Lord Kelvin, 167
Lorne, 65
Love Canal, 157
Luna Island, 100, 205
Lyell, 59
Maid of the Mist, 234
Maid of the Mist Landing, 151
map of power company land, 140
Mariposa Big Tree Grove, 45
McKim, 187
McKinley, 216
Milbert, 183
Mile Strip, 260, 263
military posts, 26
mills, 21, 28, 78
Model City, 157
Model Town Company, 157
Monteagle Hotel, 257
monumentalization, 190
Mooney, 145
Moran, 62
Morris, 260
movement, 45
Mowat, 64, 65, 86, 179
Muir, 47
murder, 135, 198
National Commissioner of Agriculture, 42
National Landmark, 226
national park idea, 43
national parks, 41, 199
native plant communities, 115
native plants, 113
native trees, 117
natural ecological characteristics, 126
natural effect, 115
natural regeneration, 119
New York Central, 99, 237
New York Central and Hudson River Railroad, 103

Niagara Association, 54, 82
Niagara Commission, 86
Niagara County Irrigation and Water Supply Company, 156
Niagara Falls, 137
Niagara Falls and Suspension Bridge Railway
Niagara Falls and Suspension Bridge Railway Company, 161, 165, 172, 191, 219
Niagara Falls Association, 53
Niagara Falls Gazette, 39, 173
Niagara Falls Hydraulic and Power Company, 31, 152
Niagara Falls Hydraulic Power and Manufacturing Company, 34, 77, 155, 165
Niagara Falls Museum, 272
Niagara Falls Power Company, 121
Niagara Falls Power Company, 77, 141, 149, 154, 155, 157, 178, 209, 215
Niagara Falls Water Power Company, 32
Niagara House, 37
Niagara Hydraulic Electric Company, 147, 157
Niagara Power and Development Company, 157
Niagara Reservation, 46, 97, 154, 158, 178
Niagara River, 27
Niagara River Hydraulic Tunnel, Power and Sewer Company, 155, 176
Niagara strait, 151
Niagara, Lockport and Ontario Power Company, 153, 157
Norman, 50
Norton, 54
Norway Maple, 116
Old French Landing, 126
old money, 38

Index

www.ingramcontent.com/pod-product-compliance
Lightning Source LLC
Chambersburg PA
CBHW070223190526
45169CB00001B/57